Comptes Rendus de l'Académie Internationale de Philosophie des Sciences

Tome 1

Science's Voice of Reflection

Tome 1
Science's Voice of Reflection
Gerhard Heinzmann and Benedikt Löwe, Éditeurs

Science's Voice of Reflection

Éditeurs
Gerhard Heinzmann
Benedikt Löwe

ISBN 978-1-84890-339-5

College Publications, London
Scientific Director: Dov Gabbay
Managing Director: Jane Spurr

http://www.collegepublications.co.uk

Original cover design by Laraine Welch

Table of Contents

The electronic book series *Comptes Rendus de l'Académie Internationale de Philosophie des Sciences*

This volume is the first volume published in the new publication series *Comptes Rendus de l'Académie Internationale de Philosophie des Sciences* (C.R. AIPS).

The *Académie Internationale de Philosophie des Sciences* (AIPS) was created in 1947 by Stanislas Dockx in order to reach a synthesis on fundamental questions of the philosophy of the sciences in an interdisciplinary manner. Among its early members were many famous philosophers and scientists, among them Paul Bernays, Evert Willem Beth, Józef Maria Bocheński, Niels Bohr, Émile Borel, L. E. J. Brouwer, Louis de Broglie, Albert Einstein, Ferdinand Gonseth, Edward Arthur Milne, and Hermann Weyl. The main instrument to bring together philosophers and scientists and reflect on fundamental questions were the regular conferences of the AIPS, usually resulting in a proceedings volume published with different publishers. Between 1947 and 2021, the AIPS held sixty-five of these conferences which we list in chronological order:

1947. *Problèmes philosophiques des sciences.*
Brussels, Belgium, 8–13 September 1947.

1949. *Les quanta et la vie.*
Brussels, Belgium, 28–30 April 1949.

1959. *Philosophie ouverte et expérience scientifique.*
Rome, Italy, 2–8 August 1959.

1961. *Les conditions biologiques indispensables à la liberté de l'homme.*
Leiden, The Netherlands, 5–8 April 1961.

1961. *Philosophie de la physique.*
Paris, France, 16–18 October 1961.

1962. *Information et prévision dans les différentes sciences.*
Brussels, Belgium, 3–8 September 1962.

1963. *La phénoménologie et les sciences de la nature.*
Fribourg, Switzerland, 2–4 September 1963.

1964. *Objectivité et réalité dans les différentes sciences.*
Brussels, Belgium, 7–9 September 1964.

1965. *Civilisation technique et humanisme.*
Ouchy, Switzerland, 1–8 September 1965.

1966. *Les fondements de la physique.*
Oberwolfach, Germany, 1–6 July 1966.

1967. *La symmétrie comme principe heuristique dans les différentes sciences.*
Amsterdam, The Netherlands, 1–3 September 1967.

1968. *La méthode prospective.*
Heverlee-Louvain, Belgium, 10–13 September 1968.

1969. *La méthode de la recherche, méthodologies particulières et méthodologie générale.*
Ouchy, Switzerland, 6–9 September 1969.

1970. *L'explication dans les sciences.*
Genève, Switzerland, 25–28 September 1970.

1971. *Science, philosophie et foi.*
Biel, Switzerland, 8–11 September 1971.

1972. *La condition humaine: de l'atome à l'eschatologie.*
Ghent, Belgium, 11–16 September 1972.

1973. *Science et métaphysique.*
Fribourg, Switzerland, 12–15 September 1973.

1974. *La sémantique dans les sciences.*
Rixensart, Belgium, 30 August–3 September 1974.

1975. *La recherche de systèmes.*
Santa Margherita Ligure, Italy, 18–22 July 1975.

1976. *Logique et ontologie.*
Salzburg, Austria, 21–24 September 1976.

1977. *Christian theology in the context of scientific evolution.*
New York NY, United States of America, 5–9 July 1977.

1978. *Spécificité des sciences humaines en tant que sciences.*
Trento, Italy, 4–7 May 1978.

1979. *Un siècle dans la philosophie des mathématiques.*
Orbetello, Italy, 17–21 April 1979.

1980. *Le corporel et le mental—The Mind-Body Problem.*
Genoa, Italy, 8–12 April 1980.

1981. *La nature de la vérité scientifique.*
Brussels, Belgium, 27–29 April 1981.

1982. *La responsabilité de la science.*
Castrocaro Terme, Italy, 13–16 April 1982.

1983. *La responsabilité de la science en tant que réponse aux attentes de l'homme contemporain.*
Seville, Spain, 5–8 April 1983.

1984. *La responsabilité éthique face au développement biomédical.*
Louvain-en-Woluwe, Belgium, 23–28 April 1984.

1985. *Les relations mutuelles entre la philosophie des sciences et l'histoire des sciences.*
Fribourg, Switzerland, 15–18 May 1985.

1986. *Le rôle de l'expérience dans la science.*
Heidelberg, Germany, 14–17 May 1986.

1987. *La probabilité dans les sciences.*
Vico Equense, Italy, 27–31 May 1987.

1988. *Science et métaphysique.*
Cesena, Italy, 19–21 May 1988.

1989. *The origin and the evolution of the universe and the mankind.*
Lima, Peru, 10–12 August 1989.

1990. *Science et sagesse.*
Fribourg, Switzerland, 23–25 May 1990.

1991. *Science and the intelligibility of the world.*
Uppsala, Sweden, 8–9 August 1991.

1992. *Interprétations actuelles de l'homme: philosophie, science et réligion.*
Naples, Italy, 4–6 June 1992.

1993. *Philosophy of mathematics today.*
Budapest, Hungary, 20–22 May 1993.

1994. *La science et l'hypothèse.*
Nancy, France, 12–14 May 1994.

1995. *Observability, unobservability and their impact on the issue of scientific realism.*
Parma, Italy, 24–28 May 1995.

1996. *Philosophy of biology today.*
Vigo, Spain, 15–18 May 1996.

1997. *Advances in the philosophy of technology.*
Karlsruhe, Germany, 20–24 May 1997.

1998. *Interpretation and meaning of illness.*
Milan, Italy, 21–24 May 1998.

1999. *Alternative logics: do sciences need them?*
Salzburg, Austria, 12–16 May 1999.

2000. *The problem of the unity of science.*
Copenhagen & Aarhus, Denmark, 1–3 June 2000.

2001. *Complexity and emergence.*
Bergamo, Italy, 10–12 May 2001.

2002. *Aspects philosophiques dans les sciences cognitives.*
Paris, France, 20–22 June 2002.

2003. *Scienza ed etica. Contesti axiologici della scienza.*
Lecce, Italy, 16–21 October 2003.

2004. *Operations and constructions in science.*
Erlangen, Germany, 17–18 September 2004.

2005. *Epistemology and the social.*
Tenerife, Spain, 22–26 September 2005.

2006. *Le naturel et l'artificiel dans les sciences de la vie.*
Mexico City, Mexico, 11–14 September 2006.

2007. *Le temps dans les différentes approches scientifiques.*
Cerisy, France, 4–9 October 2007.

2008. *Relations entre les sciences naturelles et les sciences humaines.*
Rovereto, Italy, 15–19 September 2008.

2009. *Evolutionism and religion.*
Florence, Italy, 19–21 November 2009.

2010. *Science and interpretation.*
Zadar, Croatia, 6–10 September 2010.

2011. *Representation and explanation in the sciences.*
Louvain-la-Neuve, Belgium, 25–29 April 2011.

2012. *The Legacy of A. M. Turing.*
Urbino, Italy, 24–27 September 2012.

2013. *Science—metaphysics—religion.*
Široki Brijeg, Bosnia-Herzegovina, 24–28 July 2013.

2014. *Pragmatism and the practical turn in philosophy of science.*
Pont-à-Mousson, France, 10–14 September 2014.

2015. *Scientific Realism: Objectivity and Truth in Science.*
A Coruña, Spain, 22–25 September 2015.

2015. *The European critical rationalism.*
Varese, Italy, 28–29 September 2015.

2016. *Mechanistic explanations, computability and complex systems.*
Dortmund, Germany, 28–30 October 2016.

2017. *Axiomatic thinking.*
Lisbon, Portugal, 11-114 October 2017.

2018. *Science's voice of reflection. The philosopher of science as part of the scientific endeavour.*
Amsterdam, The Netherlands, 4–7 September 2018.

2019. *The human being facing new biomedical technology, including the subject of big data, the role of artificial intelligence for their analysis, genome editing and personal medicine.*
Prague, Czech Republic, 11–14 August 2019.

2021. *The relevance of judgment for philosophy of science.*
Zadar, Croatia, 12–15 October 2021.

Academic publishing is currently undergoing fundamental changes. Traditional publication models that guaranteed dissemination of ideas in the past do not serve this purpose anymore. As a consequence, the AIPS decided in their General Assembly in Prague on 13 August 2019 to create an electronic publication series entitled *Comptes Rendus de l'Académie Internationale de Philosophie des Sciences* whose papers will be openly available to everyone in which papers presented at the conferences of the Academy can be published and disseminated.

This publication project is very generously supported by the *Institute for Logic, Language and Computation* at the *Universiteit van Amsterdam* which provides the online hosting of the website of the C.R. AIPS. In addition to the open access online publication, the volumes of the series can be ordered as traditional printed books via the publisher College Publications in London. We should like to thank Dov Gabbay and Jane Spurr at College Publications for their support in setting up this agreement and the new book series.

<table>
<tr><td>Cambridge</td><td>B.L.</td></tr>
<tr><td>June 2022</td><td>*Éditeur C.R. AIPS*</td></tr>
</table>

Science's voice of reflection.

The 2018 conference of the *Académie Internationale de Philosophie des Sciences* was held at the *Universiteit Amsterdam* in September 2018 under the title

Science's voice of reflection.
The philosopher of science as part of the scientific endeavour.

By virtue of its subject, its methods, and its disciplinary tradition, philosophy of science straddles the borderlines between C. P. Snow's *Two Cultures*, connected equally strongly to the learned realm of the humanities and the technological domain of the sciences and thereby linking these worlds.

Philosophers of science who engage with scientists or engineers often understand their role as that of the voice of reflection; the philosophical eagle perspective allows them to engage with those questions all too often ignored in the everyday routine of scientific practice: questions about the motivation, norms, values, methods, and limitations of the scientific enterprise.

Many modern scientific projects covering all of the disciplines in the natural, medical and engineering sciences urgently require this level of philosophical reflection: large-scale collaborative scientific projects with major impact on our world and society raise concerns about sustainability, safety, objectivity, inter-subjectivity, ethics, and the fundamental concepts underlying the scientific questions, all of which are firmly within the domain of competence of the philosopher of science.

Alas, philosophers of science are rarely if ever consulted or incorporated in the decision-making processes concerning these collaborative scientific projects. We emphatically believe that the focus on only one of C. P. Snow's two cultures is detremental to the larger goal of science as the endeavour of understanding and improving the world.

The 2018 symposium aimed to explore the possible and actual interactions of philosophy of science with the scientist's endeavour, including, e.g.,

historical studies, case studies of current collaborations between philosophers and scientists, the role of philosophy in the academic training of future
scientists, and many more topics. In the following, we present the schedule
of the conference with all presentations:

Tuesday, 4 September 2018.

17:15–17:35. Opening (Gerhard Heinzmann & Benedikt Löwe).

17:35–18:35. Sonja Smets, Amsterdam, The Netherlands: *Where logic meets the social sciences.*

From 19:00. Dinner.

Wednesday, 5 September 2018.

9:50–10:30. Gregor Schiemann, Wuppertal, Germany: *Epistemology of the LHC.*

10:30–11:10. Jean-Guy Meunier, Montréal QC, Canada: *Modeling the mind: Bridging philosophy and cognitive sciences.*

11:10–11:30. Coffee Break.

11:30–12:10. Valentin A. Bazhanov, Ulyanovsk, Russia: *The detour heuristic influence of philosophy upon science: The cases of ide-ologized science and Kant's program in neuroscience.*

12:10–12:50. Plenary discussion about the role of the philosopher of science in the natural and social sciences.

12:50–13:50. Lunch Break.

13:50–14:30. Michel Ghins, Louvain, Belgium: *Scientific realism and scientific practice.*

14:30–15:10. Jesús Zamora-Bonilla, Madrid, Spain: *Is philosophy's role to create concepts or to destroy them?*

15:10–15:40 Coffee Break.

15:40–16:20. Marco Buzzoni, Macerata, Italy: *Methodological natu-ralism or transcendental distinction? On the relationship between philosophy, science, and philosophy of science.*

16:20–17:00. Inkeri Koskinen, Helsinki, Finland: *Engaging through case studies: Can empirical philosophy of science influence the development of transdisciplinarity?*

17:00–17:20. Coffee Break.

17:20–19:20. General Assembly of the *Académie Internationale de Philosophie des Sciences.*

From 19:45. Conference Dinner.

Thursday, 6 September 2018.

9:50–10:30. Michael Detlefsen, Notre Dame IN, United States of America: *Rigor as an epistemological ideal of mathematical proof.*

10:30–11:10. Brigitte Falkenburg, Dortmund, Germany: *Some remarks on the relations of philosophy to the sciences.*

11:10–11:30. Coffee Break.

11:30–12:10. Fabio Minazzi, Varese, Italy: *Historical epistemology as a meta-reflection between science and philosophy.*

12:10–12:50. Jure Zovko, Zadar, Croatia: *Judgment as link between C. P. Snow's two cultures.*

12:50–13:50. Lunch Break.

13:50–14:30. Martin Carrier, Bielefeld, Germany: *Responsible Research & Innovation: Prospects & Obstacles.*

14:30–15:10. Alberto Cordero, New York NY, United States of America: *On the Complex Interactions between Science and Philosophy of Science.*

15:10–15:40 Coffee Break.

15:40–16:20. Elliott Sober, Madison WI, United States of America: *Philosophical Interventions in Science—a strategy and a Case Study (Parsimony)*

16:20–17:00. Closing.

The conference took place in the historical *VOC Zaal* in the *Bushuis* of the *Universiteit van Amsterdam*. The *Bushuis* dates back to the mid 16th century and was originally the gun magazine of the city of Amsterdam. In the 17th century, the Dutch East India Company (*Vereenigde Oost Indische Compagnie*; VOC) made it its headquarters by adding the *Oost-Indisch Huis* to the building complex (1606). The *VOC Zaal* was used as the meeting chamber for the *Heren XVII* (the "Lords Seventeen"), the board of the VOC elected from the shareholders. The *genius loci* and its historical reminder of the Dutch colonial past with violence and oppression in the name of progress served as an impressive reminder of the importance of grounding scientific development in the understanding and contextualisation provided by the humanities. The organisers acknowledge additional financial support from the *Institute for Logic, Language and Computation* of the *Universiteit van Amsterdam*.

That this proceedings volume appears almost four years after the symposium is due to the fact that its production fell into the phase of the creation of the new book series *Comptes Rendus de l'Académie Internationale de Philosophie des Sciences* (C.R. AIPS): the original plan was to publish the proceedings volume as a special issue of a journal, but this plan was abandoned when the opportunity to make this volume the inaugural volume of the new book series presented itself. The delays caused by this change of publication plan meant that many of the papers presented at the conference had already found some other publication venue in the meantime; as a consequence, only five of the presented papers are published in this volume: Marco Buzzoni's paper *Disunity in the philosophy of science: for and against*, Alberto Cordero's paper *Cooperation and conflict between philosophers of science and scientists*, Brigitte Falkenburg's paper *The two cultures–old and new debates on philosophy and the sciences*, Inkeri Koskinen's paper, *Engaging or not engaging with transdisciplinary research: on methodological choices in philosophical case studies*, and Fabio Minazzi's paper *Historical epistemology as a meta-reflection between science and philosophy*.

The AIPS decided that the book series C.R. AIPS will provide proceedings of the events of the AIPS as they happened and reflect the diversity among members of the AIPS as well as their disagreements. Therefore, the papers published in this volume are not formally peer-reviewed beyond the feedback that took place in the lively discussions in Amsterdam: the papers reflect the opinions of the authors alone with no editorial influence. The five papers in this volume cover a wide range of thoughts and positions on the subject of the interaction between the philosopher of science and the scientist and we hope that the readers will find the papers thought-provoking and interesting.

Nancy & Cambridge   G.H.   B.L.
June 2022

Disunity in the philosophy of science: for and against

Marco Buzzoni*

Dipartimento di Studi Umanistici—Lingue, Mediazione, Storia, Lettere, Filosofia, Università di Macerata, via Garibaldi 20, 62100 Macerata, Italy

E-mail: `marco.buzzoni@unimc.it`; `buzzoni@mailbox.org`

Abstract. Relatively few scholars have explicitly denied the advisability, or even the necessity of a close synergy or cooperation between scientists and philosophers, but if this is to go beyond a simple statement without philosophical justification, it is necessary to highlight the logical-epistemological roots of the complementarity of science and philosophy. Elsewhere, starting from a particular conception of the Kantian *a priori*, I have argued for a new position that draws a distinction between philosophy and the sciences in a way that relates them to one another such that they not only can, but must, cooperate. In this paper I shall explore the implications of this position for the *dis*unity of science.

In spite of some fundamental points of agreement between the disunity approach and the position sketched here, there is at least one fundamental difference concerning the relationship between philosophy and the sciences. By removing all material content (even any contingent material content) from the Kantian concept of *a priori*, the main idea of the disunity thesis is coherently defensible. My conception of the Kantian *a priori* explains philosophy's unlimited openness to any subject-matter, while placing both scientific and philosophical discourse in an inter- and intra-disciplinary dialogue: the unlimited openness of philosophy goes beyond the limits of any scientific discipline or any particular philosophical discourse, and may serve as a universal medium for the attainment of a common agreement that must be assumed as possible in principle. From this point of view, it is possible both to accept, in a qualified sense, the positivist demand for unity tacitly expressed by many objections against the disunity thesis and, at the same time, the legitimacy of an opponent who denies the central thesis of the disunity approach.

1 Introduction

Looking back, the claim that science is disunified was already present in such works as Thomas S. Kuhn's *The Structure of Scientific Revolutions* (especially in the postscript to the second edition, 1970) and Paul Feyerabend's

---

*I presented an earlier (and much briefer) version of this paper at the conference of the *Académie Internationale de Philosophie des Sciences* at the University of Amsterdam, The Netherlands (4–7 September 2018). I thank all those who contributed to the discussion of the paper during and after the conference. Special thanks to Mike Stuart, who read a draft of this article and provided helpful comments and suggestions. This work is part of the research programme submitted to the Italian Ministry of University and Research (PRIN 2020 program "Epistemology and Cognition. Theory, formalisms, and applications", Prot. 2020BYMCK9).

Against Method (1970). It is expressed much more explicitly in Jerry Fodor's paper "Special Sciences (Or: The Disunity of Science As a Working Hypothesis)" (1974), Patrick Suppes's paper "Plurality of Science" (1978), Ian Hacking's *Representing and Intervening* (1983), and Nancy Cartwright's *How the Laws of Physics Lie* (1983). The position has since been fully developed in works representative of this trend such as John Dupré's *The Disorder of Things* (1993), the essays collected in Peter Galison and David Stump's *The Disunity of Science* (1996), and Nancy Cartwright's *The Dappled World* (1999).

For our purposes, we may define "the disunity thesis" as the combination of two theses listed by Kellert, Longino, and Waters (cf. 2006, p. vii): (1) natural or cultural phenomena cannot be fully investigated and/or explained by a single theory or a single approach; (2) irreducible pluralism and disunity are not only to be found within science but also at the metalinguistic level, in the philosophies of science: scientific standpoints, methods and practices are too different to permit to suppose they may be explained by only one theory of science.

This article aims to assess the strengths and weaknesses of the disunity thesis. The critical literature has highlighted, albeit not entirely clearly and convincingly, some weaknesses of this position (see e.g., Davies 1996, Fuller 2002, Kellert et al. (eds) 2006; Ruphy 2016; Breitenbach and Yoon Choi 2017). It is important to clarify the scope and limitations of these objections, as they continue to hold the strengths of the disunity thesis hostage.

It is clear that the concept of disunity falls within the scope of philosophy of science, whether or not we accept a qualitative distinction between philosophy and science. This only apparently trivial fact implies, among other things, that the epistemological and methodological status of the concept of disunity cannot be fully understood unless the epistemological and methodological status of the philosophy of science is clarified first. If this concept is a concept of the philosophy of science, the clarification of its epistemological status presupposes, as a necessary condition, the more general clarification of the status of the discipline of which it is a particular exemplification. This clarification, in turn, depends on the relation between the two concepts that constitute "philosophy of science" as a particular philosophical discipline. For this reason, it is necessary to examine the concept of disunity in a much broader context than has done so far. As we shall see, the concept of disunity can be coherently defended only to the extent that we have first clarified the relationship between philosophy and science.

Few authors have explicitly denied the fruitfulness or even the necessity of a close cooperation between science and philosophy. The overwhelming majority of authors have in fact implicitly accepted Sellars's statement that we should not confound "the sound idea that philosophy is not science with the mistaken idea that philosophy is independent of science" (Sellars 1956, p. 301). If this is not to remain a mere statement without evidence, it is necessary to provide a justification for both the distinction and the need for cooperation between science and philosophy.

Elsewhere, starting from a conception of the Kantian *a priori* as purely functional (not material, though universal and necessary), I have tried to defend a position according to which there is a distinction between philosophy and the sciences that relates them to one another in such a way that they not only can, but must, cooperate. This reconciles the thesis of a principled difference between science and philosophy with a methodological naturalism according to which, to use Kant's words, "everything in natural science must be explained naturally" (AA, VIII, pp. 155–184: 178, lines 11–13).[1] The first part of §2 will briefly describe this position: on the one side, concerning its form, philosophy reverses the usual direction and attitude of empirical knowledge; on the other, and concerning its content, philosophy cannot arise from the void of pure analysis; it depends entirely for its content on considerations 'from the outside'—that is, from the empirical sciences and common sense. The minimal epistemological universality and normativity of philosophy required here can avoid both the illusion that philosophy possesses concepts with determined content independently of special disciplines and common sense, and the scientific natural attitude to believe that the concepts of philosophy have a broader applicability than they actually have.

Against the background of this relationship between science and philosophy, §3 will outline the scope and limitations of the disunity approach. That philosophy does not have an object of its own, and therefore must take it from disciplines that investigate reality from a variety of perspectives and at a variety of levels of organization (not predetermined *a priori*), is indeed in accordance with much recent work done under the banner of "the disunity of science". A first fundamental point of agreement with work that emphasizes disunity consists in the fact that the unity of the sciences cannot be grounded in the unity of empirical reality, especially because, as was already clear in Weber's pluralism and perspectivalism, empirical sciences can only explore reality from particular points of view, which select particular aspects of reality, relegating others to the background. A second fundamental point of agreement is that, from the point of view defended in

---

[1]Cf. Buzzoni (2019) and (2021). Kant's works are cited according to the Academy Edition, though in the case of The Critique of Pure Reason I first give the original pagination of the 1787 (B) edition published by Meiner in 1998 and edited by Jens Timmermann. In this last case, quotations are from Kemp Smith's 1929 translation, if necessary revised in the light of the Paul Guyer and Allen W. Wood's 1998 Cambridge edition.

this paper, the unity of the sciences cannot consist in one particular method or set of methods.

However, there is at least one fundamental point of divergence which is intimately connected with the relationship between philosophy and science. In order to have a coherent concept of disunity it is necessary to accurately distinguish, and at the same time to relate to each other, two meanings of 'disunity' and 'unity', one philosophical, the other scientific. The problem lies in the following antinomy. On the one hand, we have to avoid the untenable positivist conception of the unity of science, rightly rejected by the disunity theorists; on the other hand, however, in some sense we need the concept of a universal medium which the logical empiricists used to guarantee the intersubjective value of both the dialogue between the special sciences and that between the sciences and philosophy. As I will try to show, this antinomy can be resolved by rethinking Kant's definition of philosophy—according to which philosophy is occupied not so much with objects as with the mode of our knowledge of objects in so far as this mode is possible *a priori*—in the light of a complete and consistent rejection of the *material* character of the Kantian *a priori*. In a way, this is nothing new. It was the logical empiricists and all the major exponents of the tradition of the philosophy of science who most strongly expressed this rejection of a material *a priori*, even though they—like today's advocates of the contingent and relativized *a priori*—did not realize that, unlike a material *a priori*, a truly formal *a priori* is not only compatible with the conceptual changes that had transformed the physics of their time, but, contrary to their demand for a science unified in method and language, requires the necessary limitedness and disunity of the various empirical discourses aimed at exploring what we call empirical reality. The concept of empirical reality expresses only the formal or, to use Kant's term, the transcendental unity of human reason, the possibility in principle of always being able to find an agreement between those who disagree, no matter how different the assumptions from which they begin. Only in this purely formal sense, empty of any particular empirical content, is it possible to affirm without contradiction the qualitatively different status of philosophy, which goes beyond the limits of any particular science and which precisely for this reason stands above all parties, and may serve as a universal medium in a discussion able to reach a principled agreement. This, it seems to me, is the only way in which we can save the element of truth contained in the neo-positivistic idea of a unified science. As I shall try to show, the idea of a purely functional *a priori*, emptied of any material content (even of any contingent material content), is not at all in contrast with the main idea of the disunity approach. On the contrary, it seems to me the only way to make it coherently defensible.

2 Methodological naturalism or transcendental distinction? On the relationship between philosophy and science

It might be useful to distinguish two opposing conceptions of the relationship between science and philosophy. According to one of them, philosophy and science are assumed to be, ultimately, identical. All old and new versions of positivism held in different ways such a position, and today it is maintained by most forms of naturalism and experimental philosophy. In all cases, both the methods and the purposes of philosophy and science are regarded as identical. The only difference usually admitted is that the particular sciences, consisting of truths more or less separated, are not able to operate their integration or, at least, to have an overall view of them. Integrating different scientific worldviews to obtain a more general view is the task of philosophy: philosophy is in a certain sense co-extensive with all fields of scientific knowledge, and for this reason it is in a position to unify and co-ordinate the results of the particular sciences, with the purpose of attaining a very general knowledge or an overall system of classification.

Let us illustrate this position with some concrete examples. According to Herbert Spencer the sciences ignore the knowledge constituted by the "fusion" of "all the contributions into a whole", which is precisely the task of philosophy to achieve. In all this there is no discontinuity of principle, but an essential continuity between science and philosophy. According to Spencer, philosophy is a "knowledge of the highest degree of generality", which groups sequences among phenomena into generalizations of a simple or low order, and "rises gradually to higher and more extended generaliza-tions" (cf. Spencer 1888, § 37, pp. 131–132). From this point of view, the method of philosophy is the same as that of the sciences, since philosophy takes as its point of departure the widest scientific generalisations in order to "comprehend and consolidate" them up to "the highest degree of gener-ality" (a very similar conception can obviously be found in Comte, *Cours de philosophie positive*, lect. 2a, § 3). It is interesting to note that, while specifying this concept of philosophy, Spencer also touches on the prob-lem of the unity or disunity of science: "Knowledge of the lowest kind is un-unified knowledge; Science is partially-unified knowledge; Philosophy is completely-unified knowledge." (Spencer 1888, p. 134) In fact, as we shall see, the two problems, that of the relation between science and philosophy and the theme of the unity or disunity of science, are intimately connected.

As already mentioned, many forms of today's naturalism or experimen-tal philosophy have adhered to a similar view, which was mediated to the current debate in the philosophy of science by Ernst Mach (cf. Mach 1906, pp. vii–viii & 2–3) and main exponents of logical empiricism. These latter made a huge effort to bring together the domains of empirical science and philosophy that were deemed meaningful, excluding 'nonsensical' (*unsinnig*) metaphysical discourse from the realm of authentic knowledge. Philosophy has neither a particular domain of objects of its own, comparable to the subject matters of the various particular sciences, nor a method distinct from that of science: "philosophy—as Carnap famously said—can no longer be accepted as a field of knowledge in its own right, at the same level of, or superior to, the empirical sciences." (Carnap 1930–1931, p. 12; cf. also Carnap 1931, pp. 239–240)

The current use of the term 'naturalism', however, is due especially to Willard van Orman Quine, according to whom "philosophy [...], as an effort to get clearer on things, is not to be distinguished in essential points of purpose and method from good and bad science" (Quine 1960, p. 3). Both the continuity between science and philosophy and the characterisation of philosophy that we found in Spencer are taken up in the following passage from Quine, which is one of the most balanced expressions of his naturalism:

> Philosophy [...] is continuous with science. It is a wing of science in which aspects of method are examined more deeply, or in a wider perspective than elsewhere. It is also a wing in which the objectives of a science receive more than average scrutiny, and the significance of the results receives special appreciation. [...] The relation between philosophy and science is not best seen even in terms of give and take. Philosophy, or what appeals to me under that head, is an aspect of science. (Quine 1970, pp. 3–4).

The most recent defence of this viewpoint has come from many exponents of experimental philosophy (cf., e.g., Haug (ed.) 2014, and Fischer and Collins (eds.) 2015a, Sytsma and Buckwalter (eds) 2016, to which I would add at least Thagard 2010, 2014, and Ludwig 2018). Although experimental philosophy is a complex movement, which includes different philosophical currents, Goldman rightly, though *en passant*, noted that experimental philosophers are "a subclass of philosophical naturalists who have raised objections to the epistemic credentials of intuitions" (Goldman 2013, p. 12). In fact, a peculiar contribution to philosophy by experimental (and naturalistic) philosophy lies in having called attention to the fact that the use of intuitions, in science as well as in philosophy, is vulnerable to many kinds of error, and that by conducting and considering laboratory work, we can make progress towards determining the limits and conditions of proper application of our intuitions (cf., e.g., Fischer and Collins (eds.) 2015b, p. 4).

According to another and opposite conception of their relationship, philosophy is qualitatively different from science, since it has not only a domain, but also methods and problems of its own, alien to the particular sciences. Well-known is the Hegelian thesis, clearly expressed at the beginning of the

Phenomenology of Spirit, which ascribes to philosophy alone the capability to construct both its own method and its own object or content (cf. Hegel 1977, § 1), that is, on reflection, the capability to decide issues about experience without resorting to experience. As Hegel says elsewhere, the dialectic, as the law of necessary development of thought and reality, produces and conceives from itself its "positive content and outcome" (Hegel 2008, § 31). There is hardly any author today who would (explicitly) defend such an extreme point of view, and the discussion has rather focused on particular aspects of the status of armchair philosophy, with a particular emphasis on the possibility and limits of intuition and thought experiments in philosophy (some of the most important recent articles on the subject are collected in DePaul and Ramsey 1998 and Booth and Rowbottom (eds) 2014, to which I would add at least Brown 1991[2011], 2007, 2012; BonJour 1998, Williamson 2007, 2009; Chapman et al. 2013).

With respect to the opposition just outlined, we shall attempt to argue in favour of an intermediate position, as follows. On the one hand, there is a transcendental and principled distinction between the sciences and philosophy. This is in clear opposition to the naturalistic programme and experimental philosophy, at least insofar as the latter rejects both the qualitative distinction between philosophy and the sciences and the cognitive value in principle of philosophical discourse, that is, as one might perhaps say, to the extent that they are accompanied by ontological or metaphysical, and not only methodological, considerations.[2]

On the other hand, the position defended here is in accordance with the naturalistic attitude of experimental philosophy insofar as it rightly insists on the impossibility of disregarding the so-called principle of empiricism, according to which observation and experiment are the only sources of evidence relevant for the acceptance or rejection of empirical statements. This principle is often ascribed to John S. Mill (1863, p. 51) or, more recently, Karl R. Popper (1969, p. 54), but it is not without significance for our purpose that, as we have seen, it was already formulated by Kant. For the purpose of sketching an intermediate position, Kant deserves credit for having attempted to draw a qualitative distinction between philosophy and science that gives us an important hint as to how to relate them to one another in such a way that they not only can, but must, cooperate.

According to Kant, in asking what the nature and conditions of the possibility of knowledge are, philosophy "is occupied not so much with objects as with the mode of our knowledge of objects in so far as this mode of knowledge is to be possible *a priori*" (Kant KrV B 25, AA III, 43, lines 2–4).

---

[2]For more details on this point, cf. Stuart 2014 and Buzzoni 2019. For the distinction between metaphysical (or ontological) and methodological naturalism, cf., e.g., Papineau 2016.

Before seeing what consequences derive from this Kantian notion of the
relationship between philosophy and science, and therefore, both for philosophy of science and for the concept of disunity, it is first necessary to
mention and dissolve, albeit by very brief remarks, an ambiguity in Kant's
conception of the *a priori*. This will be done by presenting two developments of Kant's idea: my own Kantian conception of the *a priori*, and the
one that dominates almost unchallenged in the epistemological landscape
today.[3]

As already mentioned in the introduction, all the principal exponents
of the philosophy of science since the birth of the discipline at the end of
the nineteenth century criticized Kant for having subscribed to a view of
the *a priori* that, using Schlick's (and Husserl's) expression, was "material"
(see Schlick 1932). According to this view, the *a priori* possesses particular contents (such as those expressed by the laws of the conservation of
matter, the law of inertia, or the equality of action and reaction) that are
unresponsive to critical revision by experience. All the principal exponents
of the philosophy of science pointed out that this notion of the *a priori* was
confuted by the history of science: relativistic physics, quantum physics,
and non-Euclidean geometry had demonstrated that there are no *a priori*
principles endowed with particular contents and that are immune from revision by experience or from the adoption of different conventions (see, e.g.,
Mach 1933, pp. 458–459, Poincaré 1902 [2018], pp. 64–55 [p. 42], Reichenbach 1920, pp. 1–5, Bridgman 1927, pp. 3–9, Lewis 1929, Popper 1935, p.
188, Dewey 1938).

On reflection, what all these authors rejected was just the claim that the
Kantian *a priori* is universal and necessary, while they did not reject the
material character of the Kantian *a priori*. Strictly speaking, the material
character of the *a priori* was accepted, though in a contingent and relativized form, and it is precisely the latter form that is today defended by
almost all those who accept the usefulness of some concept of the *a priori*
in connection with Kant. This is also behind the idea of Thomas Kuhn
being "Kant on wheels" (cf. Lipton 2003), but the most important defender
and populariser of this idea, among the recent authors, is Michael Friedman
(1992, pp. 4 and 58, and 2013, p 25; as far as thought experiments are
concerned, see Fehige 2012 and 2013).

I have elsewhere argued against this account of the *a priori*, both for
reasons of historical-philological accuracy and for reasons to do with what we
want from a theory of the *a priori* (see, respectively, Buzzoni 2013 and 2005;
on the more general implications of the view of the *a priori* for the concept
of thought experiment, see Buzzoni 2018). Here, for reasons of economy,
I shall omit this line of argument and confine myself to a few implications

---

[3]The rest of this section is largely based on Buzzoni (2019) and (2021).

of this view, which are relevant both for the relationship between science and philosophy and the notion of philosophy of science. It is precisely by rejecting the idea that the *a priori* has any material content at all that I shall understand the Kantian definition of philosophy mentioned above. In this way, to state again the fundamental thesis of the paper, I shall try to reconcile unity and disunity as two apparently opposing, but in fact both necessary, aspects of how knowledge and cultural ideas change over time. On the other hand, the very idea of an *a priori* liberated from all contingent content allows us to admit the irreducible plurality of particular cultural discourses, without denying the unity (in principle) of the discourse that seeks to reconstruct (and de facto reconstructs as far as it can) the many aspects of reality in a cultural unity in constant flux. From this point of view, not only can the unifying role of philosophy be understood in a way that is in perfect accordance with the disunity thesis, but also in a way that makes both of these concepts, the unifying task of philosophy and the disunity thesis, coherently defensible.

Let us start again from the Kantian definition of philosophy, according to which philosophy "is occupied not so much with objects as with the mode of our knowledge of objects in so far as this mode of knowledge is to be possible *a priori*". Kant's claim may be expressed by saying that what is distinctive of philosophy is the fact that it reverses the direction or attitude adopted towards reality that is characteristic of scientific inquiry. Here lies the most important qualitative distinction between science and philosophy. Instead of exploring some particular aspects of natural or cultural reality, the philosopher investigates our relation to them, that is, in Kant's parlance, the conditions of the possibility of the human faculty of knowing (and morally evaluating, an aspect that will not be covered here) natural or cultural reality.

On reflection, it follows from this that each scientific (sub)discipline, since its characteristic concepts are bound to a particular point of view, has no means to answer questions about the nature and conditions of its own kind of knowledge. For this reason, it is not physics that can answer the question of what the nature and conditions of the knowledge in physics is, nor sociology for sociology; on the contrary, philosophy is not only capable of investigating the natural limits and conditions of the possibility of any other cognitive activity, but can summon itself for judgment before its own tribunal and try to clarify its own status: it makes perfect sense to speak of a meta-philosophy understood as a philosophy of philosophy.

In other words, while science is intrinsically constituted by a conscious restriction of the field of research, led by this or that particular point of view, there is nothing that can be excluded from philosophical critique. The unlimited openness of philosophy would only give rise to a futile attempt to exhaust the universe if it were not for its direction of inquiry, which is the reverse of the empirical-scientific point of view. This reversal is, in the last analysis, the deepest root of the unlimited scope of philosophy *a parte objecti*, that is, of its ability to reflect and question any kind of experience (including philosophical ones).[4]

Now, it is important to emphasise that the condition of possibility of all this can only be a purely functional *a priori*, freed of any particular content. A purely functional *a priori* does not enjoin or forbid any particular content from philosophical reflection. On the contrary, the assumption of a material *a priori* (such as the one that delimits from time to time, in a contingent and historically changing way, the field of investigation of the sciences) cannot explain the unlimited openness of philosophy with regard to its possible objects. A material *a priori* can lead us to investigate only certain contents and not others, functioning as a kind of blinkering device, which allows us to see some things and not others, depending on the cone of light that it projects on a particular area of reality rather than on another.

In order to avoid serious misunderstandings, it is important to point out that, in the perspective assumed here, a material and contingent *a priori* in no way makes knowledge impossible. It is not in contrast with the capacity to learn from experience. On the contrary, it plays a fundamental role in the typical way in which empirical-experimental knowledge proceeds: the continuous interaction between our body (or between our instruments as its extensions) and the reality around us always illuminates (i.e., makes perceptible) new aspects of reality, or, to continue with the metaphor already used, expands the cone of light that the experimental interaction projects onto reality. Empirical sciences explore reality from particular points of view, which select particular contents, and necessarily neglect others: a mechanical phenomenon results from considering reality from a partial point of view, which takes into account only some properties of reality, such as force, mass and certain spatial and temporal relations. However, by reversing the direction or attitude adopted towards reality by scientific inquiry, a purely formal *a priori* allows philosophy to get a mode of inquiry or critical attitude so generalized as to act without special or particular *a priori* limits *a parte objecti*, i.e., with respect to its possible subject-matter. In short, what makes knowledge impossible is not the assumption of a contingent and relativized material *a priori*, but the fact of not admitting at the same time

---

[4]In order to answer the question (raised by Mike Stuart, whom I thank for this) about the meaning of the expression 'philosophical experiences' in this paper, we have to note that the individual philosopher in the flesh can only practice the unlimited critique proper to philosophical discourse in the first person and from a particular perspective, determined both by our personal history (the conclusions we have come to, the decisions we have made, etc.) and by the more general histories of the social groups that, from a certain moment onwards, have interacted with our personal history.

a purely formal *a priori*, which, unlike the contingent and material *a priori*, enjoys a universality and necessity similar in principle to that which Kant had already attributed to it.

On the other hand, however, the unlimited character of philosophical reflection, which I have so far placed in contrast to the limited and circumscribed character of the empirical sciences, is only one side of the coin, the other being a limitation. Strictly speaking, to say that there is no limit whatever to the possible objects of philosophy is to say that it has no object at all. This is hardly surprising: the rejection of any material content of the *a priori* leaves philosophy no domain of objects of its own, philosophy must find its object outside itself, that is, in the natural and human sciences (as well as in the humanistic disciplines and in everyday life).

Given the purely functional nature of the *a priori*, philosophy, on the one side, and the other expressions of human culture, on the other side, though distinct in principle, are so to speak designed to cooperate, since they are supplementary and inseparable. Philosophy could not break its connection with the rest of culture without cancelling itself. It is obliged by its very nature to open itself to what is different from itself, that is to say, to what lies outside it, to the different particular fields of human life, from which it draws its (material) contents.

But what about the particular methods? One may be inclined to think that the fact that philosophy reverses the attitude or direction of scientific inquiry entails important methodological differences between philosophy and science. This reversal is indeed intimately connected with the only difference we have to concede between philosophy and the empirical sciences. These latter fulfil the requirement, proper to all rational discourse, to testify as to how things really are, by means of the construction and functioning of what I would call an 'experimental machine' (or perhaps, expressed in the more fashionable terms of today, an "experimental mechanism"), which concretely exemplifies the theoretical content of a claim about nature and its laws.

Now, recourse to experiment is only indirectly possible for philosophy, which can have access to the contents of experience only through the various sciences, common sense, and other disciplines. However, the rejection of the material nature of the *a priori*, if consistently carried out, implies that, except for the opposite direction or attitude towards reality, there is no particular method or form of reasoning that is peculiar or restricted to philosophy.

Any project to find a particular philosophical method that can completely erase this difference between science and philosophy is doomed to failure. It would be tantamount to repeating Kant's mistake of seeking a method that could put philosophy on "the secure path of a science". And this, upon closer scrutiny, is only another unfortunate consequence, or, better, another residue of a material account of Kant's *a priori*, the same seed, in fact, from which sprouted the myth of a re-foundation of philosophy *de novo* and *ab imis*, from Descartes to Husserl, from Comte to the neo-positivist idea of a "scientific philosophy". If on the one hand the logical empiricists had reduced (genuine) philosophy to science, Husserl reduced the various sciences to philosophy. He pointed out (with approval) that the Cartesian project of a radical grounding of philosophy on absolute foundations implies "a corresponding reformation of all the sciences," which "are only non-self-sufficient members of the one universal science [*unselbständige Glieder der einen universalen Wissenschaft*], that is, of philosophy itself (Husserl 1950, § 1. English transl. slightly modified). The great differences between these philosophers or movements do not detract from the fact that, although the starting point changes, the result is the same: the distinction between philosophy and science is erased, no matter whether the former is absorbed into the latter or the latter into the former.

Apart from the direction of inquiry (and the connected use of experiment), any other difference between science and philosophy can only arise from differences in the subject-matter dealt with, not from particular features that we might decide *a priori*. Thus, there is no particular set of methods that could be called philosophical without fear of being contradicted by someone who could show their use in scientific fields (Kant himself pointed out a remarkable methodical analogy between transcendental argumentation and chemical investigation: cf. KrV, B XX–XXI fn.; AA, III, 14 fn.). Feyerabend's thesis that there is no one method that can be said to have always led to success in the natural sciences, not only applies a fortiori to philosophy, but depends in the last analysis upon a purely functional view of the *a priori*. Philosophers, too, use all the methods they are capable of devising to solve their concrete problems, and they all have only one feature in common, which cannot erase their irreducible diversity and multiplicity: not only that of trying to bring to light the internal contradictions in our discourses (the mere quest to eliminate contradictions would certainly not suffice to circumscribe 'the' philosophical method: even scientists always try to eliminate the internal inconsistencies in their own discourses), but the fact that this is done after reversing the direction of empirical-scientific investigation.[5]

---

[5]This in no way excludes that, as Stuart (2015) has rightly argued, everyday linguistic interpretation too is experimental in nature. Indeed, everyday linguistic interpretation proceeds by trial and error thanks to the feedback of experience (even if it is the experience of a human science, and not of a natural science).

3   Unity and disunity in science and philosophy

Our thesis that philosophy does not have an object of its own, but must draw its content from disciplines that investigate reality from a variety of perspectives (not predetermined *a priori*), seems to be perfectly congenial to much recent work under the banner of "the disunity of science". However, in spite of important connections, there is also a relevant difference with respect to the nature of philosophy.

As we have seen, an important aspect of the relationship between philosophy and science lies in the fact that philosophy seems to play the function of relating, coordinating, classifying, bringing together or unifying the knowledge provided by the sciences. So far, I have illustrated this aspect of the relationship between philosophy and science mainly with reference to the positivistic, naturalistic, or experimentalist tendency. However, philosophers of the most different and opposite views have accepted this point: to add only one name to that of Husserl, we should mention Dilthey (Dilthey 1883, p. 146–147, English transl. 165–166). And this makes more pressing the need to properly understand this aspect of the relationship between philosophy and the sciences, which is also the central problem with the disunity thesis.

Although it is possible to draw fine distinctions within the work of those who espouse the disunity thesis, for our purposes we will focus on the two main theses, given earlier, of Kellert, Longino, and Waters (cf. 2006, p. vii):

(1) natural or cultural phenomena cannot be fully investigated and/or explained by a single theory or a single approach;

(2) irreducible pluralism and disunity are not only to be found within science but also at the metalinguistic level, in the philosophies of science: scientific standpoints, methods and practices are too different to permit to suppose they may be explained by only one theory of science.

At least at this general level of comparison, we may note some fundamental points of agreement with the position sketched above. The first of these is to be found in Weber's pluralism and perspectivalism (see Weber 1904 [1949]), which I have essentially accepted when pointing out that empirical sciences can only explore reality from particular points of view, which select particular aspects of reality, putting others in the background. Viewed from this perspective, the unity of the sciences cannot be grounded in the unity of empirical reality. Each particular inquiry searches for those particular methods that lead to the solution of particular problems. As Suppes (1978) rightly observed, mathematics (not just the empirical sciences) "is made up of many different subdisciplines, each going its own way and each primarily sensitive to the nuances of its own subject matter." (Suppes 1978, p. 8)

Thus, the sum-total of all scientific activities is not only a plurality, but also an entity that is always *becoming*, whose real existence is largely dependent upon our continuous re-thinking and re-appropriating in the first person the methods or procedures of which past research is made up.

The second fundamental point of agreement is that, in the light of what I have been saying in the preceding section, the unity of the sciences cannot consist in one particular method or set of methods. As already noted, this is a fundamental point in which one should agree with Feyerabend's anarchistic theory of knowledge: there can be no general rule or method which is in all circumstances an infallible guide to knowledge or progress.[6] In particular, the unity of the sciences cannot consist in their particular experimental methods (see on this point Suppes 1978, p. 8), unless by experimental method one means simply the very general principle of empiricism already mentioned, according to which experience must serve as the ultimate criterion about claims concerning experience.

A third important point of agreement is to be found at the epistemological and methodological level. Disunity and irreducible pluralism affect not only the sciences but also philosophy of science. Just as the different sciences explore different paths to knowledge of the world, in the same way, as historically real activities, specific philosophical views are an irreducible multiplicity. In short, it is a historically undeniable fact that disunity applies both to specific sciences and, a fortiori, to the correspondingly specific philosophies.

These points of agreement notwithstanding, there a point of disagreement which I should like to stress. It can be brought to the fore by discussing some objections against the disunity thesis. For example, Breitenbach and Yoon Choi (2017) write that one cannot give up at least an idea of unity as a regulative ideal, which takes "ongoing scientific inquiry to contribute to a single understanding of the world" and "gives us a standing reason to engage in a range of unifying activities." And even earlier, Davies (1996) had rightly noted that scientists seem act successfully "on the assumption that different branches of science can be jointly harnessed in the attempt to explain a given phenomenon." (Davies 1996, p. 9; the importance of cooperation is also highlighted by Ruphy 2016, e.g., pp. 134–135) These objections might be roughly summarized by saying that, precisely because every science is particular and specialised, the simple fact that each relates to one another and is capable of making their particular findings and (cor-

---

[6]Yet I need to forestall a possible misunderstanding: contrary to what is sometimes implicitly assumed, this is not the same as Feyerabend's slogan "Anything goes"; or, better, it is the same as Feyerabend's slogan, but only under the condition that we tacitly add: "as far as it goes". No method can be excluded *a priori*, but of course not all are always successful: only those that lead to the desired result may be regarded as good methods.

responding) methods available to each other, seems to presuppose both a single general point of view and, as a correlative element of this point of view, a single world. This seems to call into question at least those versions of the philosophy of disunity that go so far as to reject "that the plurality of accounts should be consistent" (cf. Kellert et al. 2006, p. xiv).

These objections are not entirely convincing, for they do seem to rely on a rather ambiguous notion of unity. Without the clarification of this notion, I think, it is almost inevitable to fall back, implicitly and sometimes even explicitly, either into the untenable positivistic concept of the unity of scientific knowledge or into an irreducible opposition of different points of view, which excludes any possible integration. The regulative ideal towards the overall unity or coherence of our discourses has, as such, two distinct but intimately related aspects: on the one hand, the unification process is directed towards an ideal limit placed outside the actual history of culture; on the other hand, unification will always be only partial and, therefore, disunity (and the possibility of inconsistency) will always be to some extent real and inevitable: we shall never be able to integrate the plurality of approaches or accounts into a single coherent narrative.

Now, these two aspects are equally essential, and it is important to be able to think of them as distinct and united at the same time. If the tension between these two aspects were removed in favour of unity (even if only as unity of method and/or language), we would fall back on the positivist thesis of unified science; if, on the other hand, renouncing the idea of a God's-eye point of view, the tension were resolved by foregrounding the impossibility of a complete equalisation between the unity assumed as the regulative ideal and the effective and always partial results of unification, we would be committed to the actual, historical incommensurability of different points of view.

The ambiguity that afflicts the notion of unity both in the positivist arguments currently in favour and in those of the theorists of disunity against the unity of science, can be brought to the fore by raising the question of how one should understand the claim that disunity and irreducible pluralism affect not only the sciences but also the philosophy of science. The ambiguity that makes unconvincing the mentioned objections against the disunity approach underlies the relationship, of unity and at the same time of distinction, between the scientific and philosophical level.

At the scientific level, unity and disunity are in a relationship of continuous interaction. On the one hand, as some authors have urged against the disunity thesis, the historical reality of science shows a demand for unification: scientists seem to be driven by the regulative ideal, so to speak, to achieve a complete coherence in their knowledge concerning experience.

On the other hand, again at the scientific level, but this time in accordance with the thesis of disunity, this can only be done provisionally, by incorporating different perspectives into wider, more inclusive perspectives, in an open-ended process. Since every science is particular and specialised, a completed actual unification is impossible, since it would be tantamount to abandoning scientific discourse altogether. Given the specialized character of scientific discourse, the regulative ideal of unification cannot consist in a process in which different theories constitute an actual (and metaphysical) unity, but only in the simple fact – which we have seen expressed by Davies (1996)—that every science can borrow ideas, findings, methods or reasons from every other science.

But things are somewhat different at the meta-level of philosophical discourse (including that of disunity theorists). More precisely, on the one hand, there are no particular differences between scientific and philosophical dynamics as regards the relationship between unity and disunity. It is true both that philosophical development, like scientific development, shows a demand for unification (philosophers too seem to be driven by the regulative ideal to eliminate incoherences) and that this unification can only be achieved provisionally, by incorporating different perspectives into wider, more inclusive perspectives, in an open-ended process.

On the other hand, however, there is at least one principled difference between scientific and philosophical dynamics. At the philosophical level, in contrast to the scientific one, we can, and indeed we should, assume a purely formal unifying function of philosophy (which has not by chance been, in one way or another, recognised by authors of the most diverse tendencies), at least for three reasons.

Firstly, in full accord with one of the fundamental theses of this paper, the unity in principle and the unifying function of philosophy directly follow from its reversing the perspective of scientific inquiry. This directly follows from the unlimited openness of philosophy to all possible objects, which is also a necessary condition of the possibility of a free interdisciplinary discussion. The concept of empirical reality is not an empirical concept. It expresses only the formal or, to use Kant's term, the transcendental unity of human reason, the possibility in principle of always being able to find an agreement between those who disagree, no matter how different their starting assumptions. Only in this purely formal sense, empty of particular empirical content, is it possible, on the one hand, to defend the neo-positivistic idea of a unified science and affirm without contradiction the unifying function of philosophy, which goes beyond the limits of any particular science and which precisely for this reason stands over all parties and may serve as a universal medium in any discussion aimed at a possible agreement. Philosophical openness to all objects is a necessary condition of the possibility of a free interdisciplinary discussion not only between philosophers belonging to different subdisciplines of philosophy, but also— at least as far as the most general assumptions of their field are concerned— between scientists of different special sciences.

Secondly, the legitimate main demand of the disunity approach to scientific discourse would be denied if it were simply extended to philosophy of science without further caveat, that is, if it were understood in the sense that there is no sense in which we may develop a general philosophical discourse about science, but only specific philosophies of science. The reason is that, on closer inspection, the very multiplicity of particular scientific discourses and philosophical perspectives could not even be conceived without assuming the possibility in principle of a unitary philosophical point of view. In reality, this multiplicity—and what could be called the 'partiality' of any knowledge, be it scientific or philosophical—may be conceived only from a unitary point of view, *implicitly* provided by a general philosophical discourse, whose idea cannot be resolved without residue into the various specific philosophies of science.

Third and finally, the thesis of disunity would not be internally consistent if it were unable to assume, even if only hypothetically, the possibility of its own falsity or, which is the same, if it were unable to explain the possibility of a hypothetical opponent (who, to take an extreme example, might deny the disunity claim and regard science as something universal and monolithic). The fundamental thesis of disunity can only be defended if it is able to accommodate without contradiction the possibility of an opponent who denies the very thesis of disunity. The position held by such a hypothetical opponent, in fact, is simply one of the many possible positions that contribute to pluralism and disunity, even if s/he denies the pluralist thesis. For this reason, disunity theorists must be able to admit the possibility of such an opponent without running into any contradiction. But this, in turn, is only possible if disunity is not merely one particular philosophical perspective among many, but is actually *the* attitude proper to philosophy, based on an *a priori* completely devoid of content and, precisely for this reason, capable of a universality and necessity that does not exclude any particular perspectives, including those which contest the truth of the thesis of disunity.

We could make the same point from a different perspective, saying that, in order to understand the relation of unity and disunity in the sciences (and, more generally, in human culture), we need to carefully distinguish two points of view, one reflexive-transcendental, and the other empirical-methodical. As far as the first is concerned, not only in the empirical sciences, but also in all concrete cultural discourses (philosophy included), all discussions are guided by the underlying assumption—which is a purely formal criterion—that some settlement of different opinions or rival interests is in principle always possible. Because of its pure formality, this criterion can work as an independent criterion for judging what is intersubjectively right.

Of course, we must immediately ask the question of how this criterion can be concretely realised. The answer to this question requires the introduction of the second point of view, the empirical-methodical one. If the assumption that some settlement of different opinions (or rival interests) is in principle always possible, is not to remain devoid of any cognitive (and practical) function, it must be expressed by means of concrete methodical procedures which make it possible to reconstruct, to re-appropriate and to evaluate in the first person the reasons why it should be accepted. Because we have no direct revelation of the truth of a statement, we are forced to find and retrace the 'paths' that led to its being accepted or rejected.

On the one hand, a truly universal standard has a regulative value within our dialogues because it does not coincide with any particular point of view and therefore can guarantee that there is always a difference between what we de facto believe and what we should believe. It stands as an ideal towards which any effort of believing tends, and keeps constantly before the mind the fact that any particular act of belief is imperfect. On the other hand, this universality, if taken alone, would turn out to be so abstract that it would be incapable of giving any concrete advice about how to evaluate different scientific or philosophical opinions. For this reason, a second condition has to be met: any universal claim, embedded in someone's beliefs or attitudes, must be translatable into propositions that describe the concrete methodical steps through which those beliefs or attitudes may be reconstructed and appropriated by others.[7]

An important consequence of this transcendental and, at the same time, methodical foundation of disunity and pluralism is that we do not have to abandon our own point of view when we are trying to understand and reconstruct in the first person the reasons for a different point of view. On the contrary, in order to understand that some opinions or practical choices differ from our own, we have to methodically reconstruct both our own reasons (logical and experimental, but also historical, moral, aesthetic, etc.)

---

[7]We note incidentally that this amounts to a decisive rejection of the separation between the context of discovery and the context of justification. There is no moment after which it is possible to totally disregard the context of discovery. Certainly, Pythagoras's Theorem can be used in a practical way without recalling the procedural steps that led to its discovery. But if someone challenged the validity of this theorem, we ought to reconsider and retrace in the first person the procedural steps that led (and still lead) to that theorem being asserted. And this is true in any field of human discourse: when we try to convince someone that something is true, good, beautiful, etc., we ought to offer 'reasons' which, in principle, can be reproduced and appropriated in the first person even by those who do not share our views.

and the reasons for holding the competing view, and then compare them. To understand other people's opinions, scientific as well as philosophical, we have to reconstruct both our opinions and their opinions. Thus, the possibility of an opponent is not only coherent, but strictly necessary to the coherence of the discourse asserting disunity in science and in the various historically existing discourses of human culture.

The resolution of the conflict between disunity and unity lies in the recognition of the validity of both, which can only be asserted without contradiction in connection with a philosophical discussion to which no definite boundary can be set, and which precisely for this reason can be considered as capable (though only in principle, of course) of resolving any conflict between the particular (scientific as well as philosophical) perspectives. In this sense, the irreducible disunity and the unity both of the specific sciences (and of their corresponding and historically existing philosophies) are correlative concepts, required for the explanation of plurality in science *and* in philosophy. The one would be inexplicable apart from the other, since the one is the reverse of the other, and to sacrifice the one would involve the sacrifice of the other. In this sense, philosophy is a critical reflection without boundaries that allows us to discuss the points of view of specific scientific and philosophical discourses by placing them in relation with each other, in a way that avoids splitting them into coexisting but mutually independent activities. From this point of view, the general (reflexive-transcendental) task of philosophy could be defined as the task, which needs to be continually taken up from the beginning, of relating in a common dialogue not only every science, but also every particular piece of knowledge, with the whole of human culture.

Conclusion

Because the concept of disunity is a concept originating in the philosophy of science, its epistemological and methodological status cannot be fully understood unless the epistemological and methodological status of the philosophy of science is clarified first. This clarification, in turn, is not possible without understanding the relation that exists between the two concepts that constitute philosophy of science as a discipline, that of philosophy and that of science. Elsewhere, starting from a conception of the Kantian *a priori* as purely functional (not material, though universal and necessary), I had argued for a position which draws a distinction between philosophy and the sciences that relates them to one another in such a way that they not only can, but must, cooperate. According to this account, philosophy has no limit whatever as far as its possible objects are concerned because, strictly speaking, it has no object of its own and must find its object outside itself, that is, in the natural and human sciences (as well as, of course, in common sense knowledge, which is their common starting point). On the one hand, by reversing the usual direction and critical attitude of empirical knowledge and agency, philosophy (and philosophy of science), unlike the empirical sciences, obtains an unlimited openness to all reality. On the other side, however, *concerning their content*, philosophical arguments depend entirely on considerations 'from outside'—i.e., from the empirical sciences and common sense: philosophy—and not only philosophy of science—cannot arise from the void of pure analysis.

Having summarised and re-proposed this point of view in § 2, § 3 critically examined the scope and limitations of the concept of disunity in science and philosophy of science. There are fundamental points of agreement between the disunity thesis and the position sketched here concerning the relationship between philosophy and the special sciences. One of these is that the unity of the sciences cannot be grounded in the unity of empirical reality, especially because empirical sciences can only explore reality from particular points of view, which select particular aspects of reality and neglect others. Another fundamental point of agreement is that, from the point of view defended in this paper, the unity of the sciences does not consist in one particular method or set of methods.

However, in order to have a coherent concept of disunity, it is worth carefully distinguishing, and at the same time relating to each other, two meanings of 'disunity' and 'unity', one reflexive-transcendental, the other empirical-methodical. The disunity approach risks becoming incoherent to the extent that it denies the peculiarity of philosophy, which is a critical-transcendental reflection without boundaries that allows us to discuss the points of view of both particular scientific and philosophical discourses by placing them in relation with each other. All this, in turn, can only be asserted without contradiction against the background of a purely functional account of the *a priori*. A purely functional conception of the Kantian *a priori*, which easily explains the unlimited openness of philosophy to any subject-matter, is also able to place both particular scientific and philosophical discourses in an inter- and intra-disciplinary dialogue: the unlimited openness of philosophy goes beyond the limits of any special science or any particular philosophical discourse, it stands over all parties and may serve as a universal medium for the attainment of common agreement. The general (reflexive-transcendental) task of philosophy, which needs to be continually taken up from the beginning, consists in concretely relating in a common dialogue not only every science, but also every particular piece of knowledge, with the whole of human culture. From this point of view, it is possible both to accept, in a qualified sense, the positivist demand for unity tacitly expressed by many objections against disunity and, at the same time, the possibility of an opponent who denies even the central thesis of the disunity approach.

References

BonJour L. 1998. In Defense of Pure Reason: A Rationalist Account of A Priori Justification. Cambridge University Press, Cambridge.

Booth A.R. and Rowbottom D.P. (eds) 2014. Intuitions. Oxford University Press, Oxford. Breitenbach A. and Choi Y. 2017. Pluralism and the Unity of Science. The Monist 100: 391–405.

Bridgman P.W. 1927. The Logic of Modern Physics. MacMillan, New York.

Brown J.R. 1991 [2011]. Laboratory of the Mind: Thought Experiments in the Natural Sciences. Routledge, London and New York; quotations are from the 2nd ed. 2011.

Brown J.R. 2007. Counter Thought Experiments. Royal Institute of Philosophy Supplement 61: 155–177.

Brown J.R. 2012. Platonism, Naturalism, and Mathematical Knowledge. Routledge, New York, London.

Buzzoni M. 2005. Kuhn und Wittgenstein: Paradigmen, Sprachspiele und Wissenschaftsgeschichte. In: F. Stadler and M. Stöltzner (eds), Zeit und Geschichte/Time and History, pp. 38–40. Ludwig Wittgenstein Gesellschaft, Kirchberg a.W.

Buzzoni M. 2013. On thought experiments and the Kantian *a priori* in the natural sciences. A reply to Yiftach. Epistemologia 36: 277–293.

Buzzoni M. 2018. Kantian Accounts of Thought Experiments. In J.R. Brown, Y. Fehige, M. Stuart (eds.). The Routledge Companion to Thought Experiments, pp. 327–341. Routledge, New York.

Buzzoni M. 2019. Thought Experiments in Philosophy: A Neo-Kantian and Experimentalist Point of View. Topoi 38:771–779.

Buzzoni M. 2021. The Janus-Faced Nature of Philosophy of Science. Eleven Theses. Axiomathes 31: 743–762.

Carnap R. 1930–1931. Die alte und die neue Logik. Erkenntnis 1: 12–26.

Carnap R. 1931. Überwindung der Metaphysik durch logische Analyse der Sprache. Erkenntnis, 2: 220–241.

Cartwright N. 1983. How the Laws of Physics Lie. Oxford University Press, Oxford.

Cartwright N. 1999. The Dappled World: A Study of the Boundaries of Science. Cambridge University Press, Cambridge.

Chapman A., Ellis A., Hanna R., Hildebrand T., Pickford H.W. 2013. In Defense of Intuitions. A New Rationalist Manifesto. Macmillan, Palgrave.

Davies D. 1996. Explanatory disunities and the unity of science, International Studies in the Philosophy of Science 10: 5–21.

DePaul M.R. and William R. 1998. The Psychology of Intuition and Its Role in Philosophical Inquiry. Rowman & Littlefield Publishers, Lanham, Boulder, New York, Oxford.

Dewey J. 1938. Logic: The Theory of Inquiry. Holt, Rinehart & Winston, New York.

Dilthey W. 1883 [1989]. Einleitung in die Geisteswissenschaften. Duncker & Humblot, Leipzig. 1990 (repr. in Gesammelte Schriften, Bd. 1., 9th ed., Teubner, Stuttgart). English Transl. by M. Neville, Introduction to the Human Sciences. Princeton University Press, Princeton, 1989.

Dupré J. 1993. The Disorder of Things: Metaphysical Foundations of the Disunity of Science. Harvard University Press, Cambridge MA.

Fehige Y. 2012. Experiments of pure reason'. Kantianism and thought experiments in science. Epistemologia 35:141–160.

Fehige Y. 2013. The relativized a priori and the laboratory of the mind. Towards a neo-Kantian account of thought experiments in science. Epistemologia 36:55–73.

Feyerabend P.K. 1970: Against method: Outline of an Anarchistic Theory of Knowledge, in M. Radner and S. Winokur (eds), Minnesota Studies in the Philosophy of Science, IV, pp. 17–130. University of Minnesota Press, Minneapolis.

Fischer E., Collins J. (eds) 2015a. Experimental philosophy, rationalism, and naturalism. Rethinking philosophical method. Routledge, London.

Fischer E., Collins J. (eds) 2015b. Introduction. In: Experimental philosophy, rationalism, and naturalism Rethinking philosophical method, pp. 2–33. Routledge, London.

Fodor J. 1974. Special Sciences (Or: The Disunity of Science As a Working Hypothesis). Synthese, 28: 97–115.

Friedman M. 1992. Kant and the exact sciences. Harvard University Press, Cambridge MA.

Friedman M. 2013. Kant's Construction of nature. A reading of the Metaphysical Foundations of Natural Science. Cambridge University Press, Cambridge.

Fuller S. 2002. "The Changing Images of Unity and Disunity in the Philosophy of Science". In: Stamhuis I.H., Koetsier T., De Pater C & Van Helden A (eds). The Changing Image of the Sciences, pp. 171–194. Springer, New York.

Galison P. and Stump D.J. (eds) 1996: The Disunity of Science: Boundaries, Contexts, and Power, Stanford University Press, Stanford.

Goldman A. 2013. Philosophical Naturalism and Intuitional Methodology. In: Albert Casullo and Joshua Thurow (eds), The A Priori in Philosophy, pp. 11–44. Oxford University Press, Oxford.

Hacking I. 1983. Representing and Intervening: Introductory Topics in the Philosophy of Natural Science. Cambridge University Press, Cambridge.

Haug M.C. (ed) 2014. Philosophical methodology: the armchair or the laboratory? Routledge & Kegan Paul, London.

Hegel G.W.F. 1977. Phenomenology of Spirit. Transl. by A.V. Miller. Oxford University Press, Oxford.

Hegel G.W.F. 2008. Philosophy of right. Transl. by T.M. Knox, rev. by S. Houlgate. Oxford University Press, Oxford.

Husserl E. 1950. Cartesianische Meditationen und Pariser Vorträge, In: Husserliana. Edmund Husserl. Gesammelte Werke, Band 1, Martinus Nijhoff, The Hague. English Transl. by D. Cairns, Cartesian Meditations. An Introduction yo Phenomenology, Martinus Nijhoff, 's-Gravenhage, 1960.

Kellert H, Longino H.E., and Waters C.K. 2006. Introduction: The Pluralist Stance. In Kellert et al. (eds) 2006, pp. vii–xxix.

Kellert H., Longino H.E., and Waters C.K. (eds) 2006. Scientific Pluralism. University of Minnesota Press, Minneapolis/London.

Kuhn, T.S. 1970. The Structure of Scientific Revolutions, 2nd ed., University of Chicago Press, Chicago.

Lewis C.I. 1929. Mind and the World Order. Dover, New York.

Lipton P. 2003. Kant on wheels. Social Epistemology: A Journal oft Knowledge, Culture and Policy, 17: 215–219.

Ludwig K. 2018. Experimental Philosophy. In J.R. Brown, Y. Fehige, M. Stuart (eds.). The Routledge Companion to Thought Experiments, pp. 385–405. Routledge, New York.

Mach E (1906) Erkenntnis und Irrtum. Skizzen zur Psychologie der Forschung, 2nd ed. Barth, Leipzig.

Mach E. 1933. Die Mechanik in ihrer Entwickelung, 9th ed., Brockhaus, Leipzig.

Mill J.S. 1863. Utilitarianism. Parker, London.

Papineau D. 2016. Naturalism. The Stanford Encyclopedia of Philosophy (Winter 2016 Edition), Zalta E.N. (ed.).

Poincaré H. 1902[2018]. La science et l'hypothèse. Flammarion, Paris. English Transl. by M. Frappier, A. Smith, and D.J. Stump, Science and Hypothesis. Bloomsbury, London.

Popper K.R. 1935. Logik der Forschung. Springer, Wien.

Popper K.R. 1969. Conjectures and Refutations, 3d ed., Routledge & Kegan Paul, London.

Quine W.V.O. 1960. Word and Object. M.I.T. Press, New York.

Quine W.V.O. 1970. Philosophy of Language. Philosophical Progress in Language Theory. In: Howard E. Kiefer and Milton K. Munitz (eds), Language, Belief, and Metaphysics, pp. 3–18. State University of New York Press, New York.

Reichenbach H. 1920. Relativitätstheorie und Erkenntnis *a priori*. Springer, Berlin.

Ruphy S. 2016. Scientific Pluralism Reconsidered. A New Approach to the (Dis)Unity of Science. University of Pittsburgh Press, Pittsburgh.

Schlick M. 1932. Gibt es ein materiales Apriori? In: Wissenschaftlicher Jahresbericht der Philosophischen Gesellschaft an der Universität zu Wien, pp. 55–65. Verlag der Philosophischen Gesellschaft, Wien.

Sellars W. 1956. Empiricism and the Philosophy of Mind. In: Feigl H. & Scriven M. (eds) Minnesota Studies in the Philosophy of Science, vol. I, pp. 253–329. University of Minnesota Press, Minneapolis.

Spencer H. 1888. First Principles, Appleton, New York.

Stuart M.T. 2014. Cognitive science and thought experiments. A refutation of Paul Thagard's skepticism. Perspectives on Science 22: 264–287.

Stuart M 2015. Philosophical Conceptual Analysis as an Experimental Method. In: Gamerschlag T, Gerland D, Osswald R & Peterson Wiebke (eds), Meaning, Frames and Conceptual Representation, pp. 267–292. Düsseldorf University Press, Düsseldorf.

Suppes P. 1978. The Plurality of Science. In: P.D. Asquith and I. Hacking (eds), PSA 1978, vol. 2, pp. 3–16. University of Chicago Press, Chicago.

Sytsma J. and Buckwalter W. (eds) 2016. A Companion to Experimental Philosophy. Wiley & Sons, Malden, Oxford.

Thagard P. 2010. The brain and the meaning of life. Princeton University Press, Princeton.

Thagard P. 2014. Thought experiments considered harmful. Perspectives on Science 22: 288–305

Weber M. 1904 [1949]. Die „Objektivität" sozialwissenschaftlicher und sozialpolitischer Erkenntnis. Archiv für Sozialwissenschaft und Sozialpolitik, 19:22–87.

Williamson T. 2007. The philosophy of philosophy. Blackwell, Oxford.

Williamson T. 2009. Replies to Ichikawa, Martin and Weinberg. Philosophical Studies 145:465–476.

Cooperation and conflict between philosophers of science and scientists

Alberto Cordero

Philosophy, Graduate Center of the City University of New York, 365 Fifth Avenue, New York, NY, 10016, United States of America

Philosophy, Queens College, City University of New York, 65-30 Kissena Boulevard, Queens, NY, 11367-1597, United States of America

E-mail: `alberto.cordero@qc.cuny.edu, acordelec@outlook.com`

Abstract. Much good science has been done without explicit help from philosophers. However, judging by past and recent interactions, philosophers of science can and do help clarify and advance ongoing scientific projects and facilitate the critical reception of scientific proposals. I consider three significant channels of interaction—two associated with collaborative projects and one with confrontation. They involve, respectively: (1) direct epistemological and ontological influences of philosophers of science qua philosophers in scientific endeavors and vice versa, (2) ethical calls to examine lines of research deemed potentially dangerous to individuals or society, and (3) efforts by senior scientists to protect students from exposure to critiques and "fruitless distractions."

1   Philosophy and science

As the empirical sciences began to break away from philosophy in the 19th century, many working scientists maintained strong intellectual links with the old discipline. Here are some examples.

(a) In the 1840s, Charles Darwin articulated his Natural Selection theory, taking guidance from William Whewell's philosophy of the inductive sciences (1847, 1858).

(b) Albert Einstein's relativity theories incorporated insights from 19th-century work on empiricism and realism (see, e.g., Galison, 2004).

(c) In the 1910s and 20s, John B. Watson sought to improve psychology's objectivity by embedding its discourse in a positivist framework. His rejection of introspection in psychology was furthered a few decades later by B. F. Skinner (Skinner, 1976).

(d) Niels Bohr's ideas about the role of measuring devices and the boundaries of theoretical domains drew from Kant's philosophy and positivism (Bohr 1934).

(e) Heisenberg's quantum mechanics expressed a robust version of empiricism. Later his interpretation of the theory shifted towards Kantian insights (Heisenberg 1939, 1952, 1961).

(f) In more recent times, Bell's investigations into the foundations of quantum mechanics explicitly revived interest in metaphysical epistemological themes in physics (see, e.g., Bell et al. 2001).

(g) In contemporary philosophy, many naturalist approaches see their goal as making science self-aware of the strengths and limitations of its findings, theories, and methods (see, e.g., Dudley Shapere 1984, Daniel Dennett 1995). More radical naturalists emphasize the growing continuity of science and philosophy of science, arguing that philosophy is not different in critical eagerness and style of argumentation from science or common knowledge (Alexander 2012, Sytsma & Livengood 2015). Biopsychology and bio-anthropology projects draw from analytic metaphysics, epistemology, and ethics (see, e.g., Dennett 1995, Baron-Cohen 2003).

Intellectual interactions such as these operate in varied and complex ways. However, one common trend is that philosophers of science generally seek to contribute results that can help scientists articulate new hypotheses—improving their internal coherence, plausibility, and compatibility with received scientific and philosophical information. Accordingly, they raise questions about the scope and limits of ongoing scientific approaches, scientific standards of evidence, motivation, and underpinning values (epistemic and non-epistemic). The resulting analyses by philosophers often gain recognition from scientists, connecting with their technical work. In recent times this is apparent in many fields, notably post-Bell physics (as reported in, e.g., Bell et al. 2001, Cordero 2019), evolutionary biology, and experimental psychology (see, e.g., Dennett 1995, Sterelny 1999, Sober & Wilson 1998, Baron-Cohen 2003), to mention just some cases.

On the other hand, some scientists consider all the noted philosophical efforts irrelevant to their practice. Limited receptiveness and even hostility to suggestions from philosophy are widespread, especially among leading physicists. Recall, for instance, the quick way Richard Feynman and his circle dealt with the interpretive problems posed by infinite integrals in perturbation theory (renormalization) in the 1940s. This neglect is also apparent in the idea that the electron can go temporally backward, among many other proposals. (Mathematically speaking, an antiparticle traveling forwards in time is indistinguishable from the corresponding particle traveling backward). It took time for philosophers of science to develop analytic projects in tune with these and other radical metaphysical proposals from quantum theory. They did it, however. In the 1970s, philosophers of physics began to offer increasingly coherent explications of the locality principle in modern physics theories, quantum non-separability, the many-worlds interpretation, multiple-times, the block universe, space-time point reality, to name a few developments. Since then, the intellectual and methodological contributions of the philosophy of science are on view in a plethora of transformative works.[1]

Nonetheless, many scientists in foundational fields don't care much about professional philosophers' insights, preferring their philosophical intuitions. Some believe that philosophy is dead—an idea Stephen Hawking endorsed in some of his final writings and public appearances (e.g., Hawking 2010). Such neglect, however, often results in ideas that, it seems, would benefit from more significant interaction with contemporary philosophers of science. Consider, for instance, the central thesis proposed in the generally delightful book *The Mathematical Universe* by Max Tegmark (2014). In it, Tegmark argues that the Universe is a "Multiverse." A Multiverse is a multi-level entity utterly big and strange, with levels described first by the standard mathematical physics, then by physics under variations of the "constants of nature", and thirdly by many-worlds quantum mechanics. Provocatively, the book claims that *all mathematical structures exist.* Exemplifying one of the contemporary roles of philosophers of science, Jeremy Butterfield has taken Tegmark's Platonist intimations to task. In a paper titled "Our Mathematical Universe?" Butterfield argues that even if one agrees that there is a mathematical multiverse, we still need to distinguish between applied mathematics (theoretical physics) and pure mathematics—the Multiverse is an *applied mathematical structure.* The claim 'There is a mathematical multiverse' holds for pure mathematics, Butterfield notes—i.e., all possible mathematical structures are equally real. However, he adds, this Platonist stance about pure mathematics has nothing to do with a physical multiverse. From the premises that (1) 'nature is an applied mathematical structure' and (2) 'there are a plethora of pure mathematical structures,' one cannot infer that 'nature is one of many equally real structures.' Tegmark, that is, commits the fallacy of equivocation. In propositions (1) and (2), 'mathematical structure' is equivocal between applied and pure structures. One can be a Platonist about pure mathematics (and believe in ever so many pure mathematical structures) and accept all this without believing that the physical Multiverse is a purely mathematical structure.

Similar interactions between philosophy of science and science are readily on view across the sciences. The point to highlight here is that philosophy is far from "dead." Contemporary philosophers of science make logical, epistemological, and ontological contributions. Furthermore, the latter seemingly help scientific investigations—and vice versa. The next section considers a complementary channel of interactions, focused on a different angle: the ethical scrutiny of scientific projects.

---

[1] Instances in point include, e.g., Albert (2003, 2013), Wallace (2012), Maudlin (2012, 2019), Lewis (2016), to mention a few contributions from just philosophers of fundamental physics.

2   On free inquiry

The ethical side of research comprises far more than justifying the allocation of resources out of finite public means available. Social and ethical issues arise when a line of inquiry touches topics of expected impact on individuals or society.

In this section, I consider the ethical scrutiny of scientific projects by philosophers of science. To make the topic manageable within the space available, I will focus on critical evaluations that oppose the conduction of specific lines of research on ethical grounds. The issue at stake is the idea of freedom of scientific investigation. I will examine several responses and suggest how arguments in progress attest to the lively engagement of philosophers of science in current debates. My focus will be on proposals that seek to articulate and clarify ethical critiques levied against specific scientific projects, also help dialogue between scientists and their critics (and the society at large).

Consider the following case of current interest. Recent evolutionary psychology theories propose that differences in cognitive performance between males and females shown by current surveys do not seem to come exclusively from cultural factors but also partly from *biological* differences (nativist explanations). To some critics, entertaining this kind of hypothesis is ethically problematic, given the possible uses and abuses that even preliminary results might have. Nativist theories about the existence of cognitive sexual differences could exacerbate ongoing injustice on specific groups—e.g., by supporting repugnant social policies and pre-existing prejudices, as has occurred repeatedly in the past. This possibility is no small fear. Human groups (particularly women and some ethnic groups) have been grossly discriminated against numerous times based on "biological" arguments that subsequently proved either seriously invalid or unsound.

So, are cognitive differences between human groups a taboo topic in enlightened society? How are research choices on the matter to be made, and by whom? Philosophers of science play a role here. One distinguished and controversial participant is Philip Kitcher, who invites us to decide in terms of the collective good that inquiry should promote in a democratic society (Kitcher 2001). His social-minded approach is especially critical of recent projects in evolutionary psychology to study alleged cognitive differences between average male and female performances. Alleged Darwinist hypotheses on such differences prompt bitter clashes (intellectual and legal) in liberal societies. The standard accounts of average academic performance variations focus strongly on local environmental factors, particularly *cultural* ones (nurturist explanations). In the social sciences, the common view is that we have become "creatures of culture" to such an extent that our evolutionary origins can tell little, if anything, about present cognitive differences between human groups. The issues at stake are numerous and deeply felt; approaching them thus calls for caution. Perhaps the most promising way to do so is to tackle calls for research censorship in this area is on a case-by-case basis.

The reactions to nativist projects in psychology open fronts of inter-action between philosophers and scientists. These can be cooperative or negative. The epistemological and methodological difficulties faced by hy-potheses about psycho-biological predisposition are numerous. For example, distinguishing between inheritance and learning from experience can be ex-ceedingly hard—inherited traits often have "maturation" periods of many years. Nevertheless, it seems that progress in handling these difficulties has been made in the last half-century (see, e.g., Baron-Cohen 2003, also, Sterelny and Griffiths 1999, Part V). More difficult to approach are the ethical difficulties associated with nativist research. Many current projects raise concerns about ethical damage that even the very act of making in-quiries explicit might cause (some thinkers claiming that even discussing certain nativist hypotheses leads to effective discrimination).

Consider, e.g., the question of why, despite so much egalitarian invest-ment in education since the 1960s, still most top young mathematicians and theoretical physicists continue to be males. The empirical correlations between gender and certain analytic skills may all be the result of cul-tural inertial forces from the past. Or the cause may be something else. One working hypothesis proposes that, because of natural selection in Pale-olithic environments, males are on average genetically both better disposed and more inclined to analytic thinking than females, particularly at the highest end of the achievement distribution. If this is correct, the found differences are part of our *Darwinian* nature. Working along these lines, evolutionary psychologist Simon Baron-Cohen (2003) argues that, overall, the female brain is more hard-wired for empathy intelligence, while the male brain is more hard-wired for analytic understanding and system building. I.e., Darwinian evolution developed men's and women's brains differently. To nurturist critics, Baron-Cohen stresses the role that evolution and genes could play in determining men's and women's brain types while playing down social and cultural influences. Nonetheless, his theory articulates var-ious consilient Whewellian-Darwinian inductions from animal studies, evo-lutionary biology, endocrinology, brain studies, and genetics. Baron-Cohen and his collaborators at Cambridge further propose that people with autism and Asperger's syndrome have an extreme version of the male brain, along with startling novel predictions regarding prospective findings of genes that control empathizing and systemizing.

Baron-Cohen's nativist project has crucial gaps to fill; it is a work in progress. It remains unclear, e.g., whether decoding the human genome will pinpoint genes that control empathizing and systemizing, as Baron-Cohen claims. Baron-Cohen's group is aware of the hurdles and moves carefully regarding empirical correlations and their interpretations on the methodological and epistemological fronts.

Critics object to this and similar projects, especially on technical and—more inflexibly—ethical grounds. A major focus of technical objections to nativist projects centers on purported causal interpretations of experimental correlations. These seem potentially damaging enough to call for permanent vigilance. Still, Darwinian psychologists and anthropologists claim to have some ways of assessing the objectivity of psycho-biological claims in crucial areas.[2] Ethical considerations can be more difficult. From the 1970s on, the whole genre to which Baron-Cohen's project belongs has been the subject of scathing objections from major scientists and philosophers, conspicuously Richard Lewontin (1975), the late S. J. Gould (1980/1989, 1981), and Philip Kitcher (1997, 2001). Their critiques are fair regarding many specific proposals. Time and again, in the last century, the general public was rushed into believing that biological investigations had revealed all sorts of "unpleasant truths" about the existence of natural differences between some human groups. The allegations were subsequently found to have been wrong—though not before doing significant damage. Thus, there are reasons to be wary of certain nativist inquiries.

The question is how far those arguments apply to nativist inquiry in general. Calls for casting moral opprobrium on nativist research inquiries have received a boost from a general consequentialist argument articulated by Kitcher. In his view, there can be no right to free inquiry in problematic fields because the prevailing social context provides enough grounds for ethically condemning the highlighted nativist inquiries very broadly (2001, Chapter 8). By the argument's terms, Baron-Cohen's project would seem to come out as unacceptable, despite its methodological and epistemological caution.

Kitcher's consequentialist argument proceeds from the following four premises regarding a human group G. Suppose that:

The low standard of living of people in group G originates, to a significant extent, from a view C erroneously held in the past as dogma. (K1)

Even though C is now officially rejected, it lingers dangerously in society, because of a strong tendency to inflate evidential support in favor of C (epistemological asymmetry). (K2)

---

[2]See, e.g., Baron-Cohen 2003, chapters 4, 6, 8, and 10.

The society in question is politically biased toward C (e.g., news of results contrary to C would not lead to any social action in favor of G). In contrast, the slightest rumor favorable to C (K3) would raise C's popular and official credibility, with damaging consequences for G (political asymmetry).

Conclusion 1. In situations where free inquiry would increase the burden on G, there can be no right to free inquiry.

Research into the truth of nativist hypotheses regarding any possible superiority in cognitive faculties between men and women   (K4) is virtually guaranteed to increase the current burden on women.

Conclusion 2. There can be no right to free inquiry into the truth of such hypotheses.

Corollary. The inquiries in question deserve moral opprobrium because far less controversial than any duty to seek the truth is the duty to care for those whose lives already go less well and to protect them against foreseeable occurrences that would further damage them (K1).

The argument just presented calls for ethical constraints on scientific research. Its assumptions are controversial and invite cooperative scrutiny from philosophers of science. Several intertwining lines of considerations call for clarification. First, are the premises compelling? Do the intended conclusions follow? Do our current social realities provide reasonable grounds for deeming the said evolutionary inquiries ethically condemnable?

Secondly, some considerations overlooked by practicing scientists and philosophers need to be made salient (Cordero 2005):

(a) Are the terrible consequences envisaged in Kitcher's consequentialist argument a likely outcome in contemporary liberal democracies? It is not in question that political agendas can co-opt scientific debates and inquiries. Prime exemplifications abound in the form of persistent discrimination against women, 'mob racism', and the phenomenon of 'Scientific Creationism', to mention a few varieties. However, as Kitcher appreciates, cases like these also attest to civil society's actual power to efficiently limit the impact of mob epistemology through legal containment. The situation is different in authoritarian societies, but there the dangers of rational forms of inquiry to human groups are negligible compared with those posed by the state.

(b) All research into the human condition is difficult and dangerous. However, it is far from clear that trying to learn about human nature from a Darwinist perspective is *more difficult or dangerous* than trying to learn about human individuals or human groups from a sociological or any other perspective. Furthermore, it seems dubious that the disadvantaged among us would benefit from discouraging any kind of *serious* research, especially in societies marred by political and epistemic asymmetries.

Kitcher's consequentialist argument depends heavily on specific context. In Darwinian conjectures about natural differences in social and psychological dispositions between men and women, two observations come to mind. First, the structure and motivation of the noted nativist hypothesis are rooted in current evolutionary biology. As such, not just any conjecture will do as a working hypothesis. There is no room for genetic determinism since biology accepts that phenotypes are shaped jointly by genes and the environment.

Furthermore, evolutionary claims about complex phenotypes are primarily about *tendencies*, and so they are compatible with virtually any given single case outcome imaginable within the relevant total range of performance. For instance, the Darwinian suggestion that members of some group *G* might be, on average, less naturally gifted than non-members for original thinking in mathematics or theoretical physics is fully compatible with the most accomplished individual in those fields being a member of group *G*. Even strong believers in a Darwinian suggestion about the male brain do not consider the outstanding mathematical talent of Amalie Emmy Noether as a counterexample to their belief. Relevantly, in Darwinist conjectures, the reference to natural tendencies is characteristically indirect in at least two ways: (1) The germane probabilities are second-order, in that they correspond to averages over probabilistic trends at the individual level; and (2) at the individual level, tendencies operate against the backdrop provided by the environment and past experience on the one hand, and the effects of most individual organisms' *ability to learn* new behaviors—to acclimatize to a new stressor (see, e.g., Dennett 1995, Chapter 3).

Even if the scientific news turned out to be very bad for some given human group, there is a solid reason to expect the findings in question to come with an array of biological and genetic pointers of theoretical and practical significance. Suppose, e.g., that it became unreasonable to scientifically deny that members of some group *G* are, on average, less naturally gifted than non-members for some celebrated aspect of human excellence. Some might hastily conclude that members of *G* should henceforth be regarded as hopeless in the specified respect, regardless of training and education. Yet, we already know this conclusion to be false. It is a *fact* that proper training can bring practically all human beings to master basic college-level mathematics and such. Nor would it be correct to conclude that individuals cannot reach high in any significant area where they rank low as a *group*, for we also know this to be false. And something else is incorrect as well, namely the intimation that our distinctly human traits are simple, one-dimensional features—they are not. These clarifications are, however, only part of the story. In contemporary natural science, beliefs are not isolated but develop in entangled clusters. As with research into oncogenes, no matter how distressing a research result might prove to be for some people, there is reason to expect that it will also point to the design of correctives— chemical, genetic, educational—to be made available to interested individuals.   The debate over the above points remains alive. My discussion here aims to suggest how philosophers of science are trying to clarify and better articulate theories like Baron-Cohen's. The suggestions above focus on the texture of theoretical belief in the contemporary natural sciences and the role of inquiry in fallibilist contexts.

The considerations outlined also seem helpful to society at large. We live immersed in scientific ideas and products like never in history, yet the average scientific literacy keeps falling in most contemporary societies. As a result, public understanding of the scope and weakness of mainstream ideas tends to be shaped more by ideology and propaganda than critical reflection. Philosophers of science can help citizens better understand the promises, limitations (both epistemic and ethical), and prospective ethical impact of scientific proposals.

In this section, I have presented an ongoing debate on the epistemology and ethics of nativist hypotheses as exemplifying an opportunity for fruitful interaction between philosophy of science, science, and contemporary liberal society. If the suggested considerations are on target, there is ample room for mutually beneficial interaction between philosophers of science and scientists. However, some scientists reject advice from philosophers on methodological grounds. The following section considers some reactions of this sort.

3   Help not always welcome

As entwined as the philosophy of science and science are, their expectations diverge at multiple levels—enough, according to some, to limit fruitful interaction between them. On one school of thought, exposure to history and philosophy of science (HPS) can be even *unhelpful* to the practice of science. In the heyday of anti-positivist critique, Thomas Kuhn (1959, 1962) and Paul Feyerabend (1974) suggested that HPS can be detrimental to working scientists because of the revisionary claims historians and philosophers often make about science. These thinkers compared theorists working at the cutting edge with athletes competing in Olympic Games, not to be bothered with subtle critical elucidations of their practice while running,

especially about how their outcomes fall short of avowed ideals of thought
and behavior. The most genial minds of science, they noted, routinely tres-
pass the received categories of understanding. Moreover, scientists do this
often as if in a state of rapture, proposing deviant, sometimes initially incoherent, approaches through which they proceed fruitfully, on the whole,
oblivious of challenges posed by historical, epistemological, or metaphysical
doubters—let alone philosophers.

Two suggestions in this negativistic view of the interface between HPS
and science are worth highlighting.

(a) Scientists, it is noted, draw strength from a progressive picture of sci-
    ence and the scientist as rational, open-minded participants. On this
    ideal picture, scientists always proceed methodically, grounded indis-
    putably in the outcome of controlled experiments, seeking objectively
    for the truth, ready to let the chips fall where they may. Contempo-
    rary historians and philosophers of science challenge this pragmatically
    fruitful professional ideal and public image. So, the argument goes,
    to the extent that HPS propounds ideas at odds with the progressive
    view, supervisors should shield scientists at the start of their careers
    from HPS.

(b) It is further claimed that writings on HPS are usually not acceptable
    for publication in leading scientific journals. So, science majors and
    graduates will likely waste their time doing work on those issues (Kuhn
    1959: 344).

The relevant point here is the claim that learning about what philosophers and historians say regarding scientists' existing standards and behavior can be "demoralizing" for aspiring students. The image under attack
presents scientists as exemplary rational, open-minded investigators. Evidence, however, suggests that scientists operate in considerably more subjective ways. Experimental verification is often of secondary importance
compared to non-standard scientific arguments (e.g., from metaphysics and
religion), at least during some of the significant conceptual changes in science. For example, while Ptolemaic astronomers faced numerous refuting
instances, for at least a century, the Copernican theorists faced arguably
even more extreme refuting cases, compounded by severe conceptual conun-
drums. According to Kuhn (1962) and other critics, heliocentrism, favored
on quasi-mystical grounds, gained strength in influential circles between
the 1540s and 1640s. Its challenges were rendered ineffective by ad hoc
hypotheses and clever techniques of persuasion.

Numerous other examples of debunking cast similarly "negative" light on
scientific discoveries. Cases in point include Copernicus, Galileo, Lavoisier,

Dalton, Mendel, and Robert A. Millikan, to mention a few. In Feyerabend's view (*Against Method*, 1974), the slogan "anything goes" summarizes the history of science. Science, he claimed, is wonderful but does not deserve any special status because it is "just" another human project among many, closer to myth than scientistic philosophy is prepared to admit. Science, Feyerabend urged, is one of the many forms of thought that our species have developed, and not necessarily the best. Like the above from the 1960s, abrasive charges continue strong to this day outside mainstream philosophy of science (notably in some "postmodernist" projects).

Most analytic philosophers rejected early on this pessimistic view of scientific education. As Israel Scheffler admonished at the start of the anti-objectivist turn, the relativist narratives purport to establish some acid anti-objectivist claims, in particular these. (a) Scientific theory "is not controlled by data, but that data is manufactured by theory. (b) We cannot evaluate rival hypotheses rationally, there being no neutral court of observational appeal nor any shared stock of meanings. (c) Scientific change is a product not of evidential appraisal and logical judgment but intuition, persuasion, and conversion. (d) Reality does not constrain the thought of the scientist but is rather itself a projection of that thought. To this Scheffler (1967) responded: "Unless the concept of responsible scientific endeavor is to be given up as a huge illusion, the challenge of this alternative must, clearly, be met."[3] Since then, further doubts have been raised against anti-objectivist, Neo-romantic approaches over the last decades. Detailed critiques by Shapere (1964, 1980, 1984), Stephen Toulmin (1972), and numerous others have challenged the historical cases invoked by Kuhn and Feyerabend—e.g., regarding the rise of Copernican astronomy and the ousting of Newtonian theory by Einstein. The traditional idea of cognitive progress was over optimistic, but it is not as naïve as Kuhn and his followers claim.

According to objectivist critics, historical and philosophical studies might challenge the brightest science students, but that can be a *good* thing, not at all counter motivational. As Stephen G. Brush argued in seminal writings on the history of science after Kuhn (e.g., Brush 1974), historians must do more than document the application of objectivity to scientific problems. They must be prepared to analyze the philosophical, psychological, and sociological aspects of scientific work, explain how specific issues came to be considered "scientific," and how particular standards happened to evaluate solutions to those problems. The historian may also have to account for scientific change in terms other than linear progress from error toward truth. Most importantly—as far as education is concerned—learning about the historical and philosophical adventure surrounding current science can be enlightening to science students. It ostensibly was to Charles Darwin, Henri

---

[3]*Science and Subjectivity*. Indianapolis: Bobbs Merrill (1967): v-vi.

Poincaré, Albert Einstein, Louis de Broglie, Niels Bohr, Erwin Schrodinger, Werner Heisenberg, and Paul Ehrenfest, to mention a few notable cases.

Still, the pragmatic objections to exposing science students to HPS may seem to stand. Too many scientists seemingly derive strength from sanguine ideas about truth and progress that—history and cold reasoning suggest— are better left unexamined critically to do good science. HPS, which seeks to enhance self-awareness, freedom, and responsibility, may not be good for everyone. The debate on this matter continues. On their part, contemporary objectivists offer an increasingly rich and nuanced view of the relationships between science and society. In recent decades, a representative of the objectivist shift is a family of projects that now go by the label "Selective Realism."[4] On this family of approaches, the most successful scientific representations of the world are not completely correct. Still, they are not totally wrong either: successful scientific theories generally contain parts that make them "approximately correct" rather than "True, Pure, and Simple."

Selective realists respond to the problems posed by post-Kuhnian antirealists. They do so particularly regarding the empirical underdetermination of theories, the availability of skeptical readings of the history of science, and postmodernist skepticism and relativism. The principal selectivist move is to drop the more extreme claims of earlier realists. According to selectivists, empirically successful theories generally turn out to be only partly correct: taxonomies of natural types have mushy boundaries; standards for assessing scientific results change along with science, fundamental ontologies can be seriously off the mark. The natural philosophies of Galileo, Newton, Darwin, Einstein, Bohr, and numerous others strongly suggest that there are no meta-scientific criteria for accepting and rejecting scientific proposals. Humans, not nature, confer scientific significance (be it of observations, test results or an entire research line). So, over time, scientific facts—what Kitcher calls 'subversive truths'—undermine deep old beliefs and value systems and allow us to operate more effectively in the world (Kitcher 1993, 2001). Having purged Scientific Realism of excessive optimism, the next selectivist task is to show how to articulate a robust and substantive realist stance about scientific theories well-grounded in exacting scientific evidence. Kitcher is optimistic about the challenge: '[T]here is no basis for believing that value judgments inevitably enter into our appraisal of which of a set of rival hypotheses (if any) is approximately correct' (Science, Truth, and Democracy, p. 41). Notably, the noted departures from traditional scientific Realism seem to allow for a strong stance (Cordero 2017). Selective

---

[4]Started in recent times by John Worrall (1989) with a structuralist focus and by Philip Kitcher (1993) with a focus on content. The general strategy was then variously developed by (Stathis Psillos (1999), Juha Saatsi (2005, 2011), Ioannis Votsis (2011), Peter Vickers (2013), Mario Alai, and Alberto Cordero (2016), among others).

Realism is still a work in progress. Still, the point here is that the nega-tivistic tensions between science and philosophy have been (and continue to be) addressed by philosophers of science, with some promising outcomes on the horizon. For the moment, at least, the "death of publicly relevant philosophy" does not seem in sight.

4 Concluding remarks

I have explored some of the roles that contemporary philosophers of science can and do play in science and the public discussion of science. Section 1 considered representative cases where practicing scientists explicitly resort to philosophy in their work. § 2 considered the ethical scrutiny of an on-going scientific project and its impact on freedom of research. I focused on a particular case from evolutionary psychology, trying to display roles played fruitfully by philosophers of science in the current debates. Sec-tion 3 considered ideas propounded in the 1960s and 1970s, still embraced in some quarters, to the effect that history and philosophy of science can be *unhelpful* to the practice of science.

My overall conclusion of the above considerations is that there are nu-merous channels of interaction between philosophers of science and scien-tists. Although I have considered only three, the tracks highlighted seem-ingly illustrate how the philosophers of science can play a fruitful part in the scientific endeavor today.

References

Albert, David Z. (1993): Quantum Mechanics and Experience. Cambridge, MA: Harvard University Press.

Albert, David Z. (2003). Time and Chance. Cambridge, MA: Harvard University Press.

Alexander, J. (2012): Experimental Philosophy: An Introduction. Malden, MA: Polity Press, 2012

Baron-Cohen, Simon (2003): The Essential Difference. Men, Women and the Extreme Male Brain. London: Allen Lane.

Bell, John S. et al. (2001): John S. Bell on the Foundations of Quantum Mechan-ics. Singapore: World Scientific Pub. Co. Inc.

Bitbol, Michele & Stefano Osnagh (2013): "Bohr's Complementarity and Kant's Epistemology." Bohr, 1913-2013, Séminaire Poincaré XVII: 145 – 166.

Bohr, Niels (1934/2011): "The Atomic Theory and the Fundamental Principles underlying the Description of Nature," in Atomic Theory and the Descrip-tion of Nature: 102-138. Cambridge: Cambridge University Press.

Brush, Stephen G. (1974): "Should the History of Science Be Rated X?" Sci-ence (Vol. 183): 1164-1172

Brush, Stephen G. (1976). The Kind of Motion We Call Heat – A History of the Kinetic Theory of Gases in the 19th Century. North Holland, 2 volumes

Chiesa, Mecca (1994). Radical Behaviorism: The Philosophy and the Science. Cambridge, MA: Center for Behavioral Studies.

Cordero, Alberto (2005): "Contemporary Nativism, Scientific Texture, and the Limits of Free Inquiry." Philosophy of Science (72): 1220-1231.

Cordero, Alberto (2011): "Scientific Realism and the Divide and Conquer Strategy." Philosophy of Science, (78): 1120–1130.

Cordero, Alberto (2019): Philosophers Look at Quantum Mechanics. Springer Nature Switzerland: Synthese Library.

Dennett, Daniel C. (1995): Darwin's Dangerous Idea. New York: Simon & Schuster. Galison, P. (2004): Einstein's Clocks, Poincare's Maps: Empires of Time. New York: W.W. Norton & Co.

Gould, Stephen J. (1980/1989): "Sociobiology and the Theory of Natural Selection". Reprint in Michel Ruse (ed.), Philosophy of Biology. London: Macmillan, 253–263.

Gould, Stephen J. (1981), The Mismeasure of Man. New York: Norton.

Hawking, Stephen (2010): The Grand Design. New York: Bantam.

Heelan, Patrick (1965): "Reality in Heisenberg's Philosophy" - Chapter Eight of Quantum Mechanics and Objectivity.

Heisenberg, Werner (1930/1943): The Physical Principles of the Quantum Theory (Translators Eckart, Carl; Hoyt, F.C.). Dover.

Heisenberg, Werner (1952/1966). Philosophic Problems of Nuclear Science. Translated by F. C. Hayes. London: Faber & Faber.

Heisenberg, Werner (1961): "Problems of Modern Physics" by W. Heisenberg in On Modern Physics, C.N. Potter London: Orion Press.

Kuhn, Thomas S. (1959/1963). "The essential tension: Tradition and innovation in scientific research," reprinted in Scientific Creativity, Its Recognition and Development, C. W. Taylor and F. Barron. Eds. New York: Wiley.

Kuhn, Thomas S. (1962/1970): The Structure of Scientific Revolutions, Chicago: University of Chicago Press (1970, 2nd edition, with postscript).

Kitcher, Philip (1993): The Advancement of Science. New York: Oxford University Press.

Kitcher, Philip (1997): The Lives to Come. New York: Touchstone.

Kitcher, Philip (2001), Science, Truth, and Democracy. New York: Oxford University Press.

Lewis, Peter J. (2016): Quantum Ontology. Oxford: Oxford University Press.

Lewontin, Richard C. (1975), "Genetic Aspects of Intelligence", Annual Review of Genetics (9): 387–405.

Maudlin, Tim (2012): Philosophy of Physics: Space and Time. Princeton, NJ: Princeton University Press.

Maudlin, Tim (2019): Philosophy of Physics: Quantum Theory. Princeton, NJ: Princeton University Press.

Psillos, Stathis (1999): Scientific Realism. London: Routledge.

Shapere, Dudley (1964): "The Structure of Scientific Revolutions," Philosophical Review (73): 383–94.

Shapere, Dudley (1980): "The Character of Scientific Change," in T. Nickles (ed.), Scientific Discovery, Logic, and Rationality. Dordrecht: D. Reidel (Boston Studies in the Philosophy of Science 56): 61-102.

Saatsi, Juha (2005). "Reconsidering the Fresnel-Maxwell Case Study." Studies in History and Philosophy of Science 36 (3): 509–38.

Saatsi, Juha and Peter Vickers (2011): "Miraculous Success? Inconsistency and Untruth in Kirchhoff's Diffraction Theory." British Journal for the Philosophy of Science 62: 29–46.

Shapere, Dudley (1984): "Objectivity, rationality, and Scientific Change," Proceedings of the Biennial Meeting of the Philosophy of Science Association 1984: 637-663.

Skinner, B.F. (1976). About Behaviorism. New York: Random House.

Sober E., Wilson D.S. (1998): Unto Others: The Evolution and Psychology of Unselfish Behavior. Cambridge, MA: Harvard Univ Press.

Sterelny, Kim, and Paul E. Griffiths (1999), Sex and Death. Chicago: University of Chicago Press.

Sytsma, J. and J. Livengood (2015): "The Theory and Practice of Experimental Philosophy." Peterborough: Broadview Press.

Tegmark, Max (2014): Our Mathematical Universe. New York: Random House.

Vickers, Peter (2013): "A Confrontation of Convergent Realism." Philosophy of Science 80: 189-211.

Votsis, Ioannis (2011): "Saving the Intuitions: Polylithic Reference." Synthese 180 (2): 121–37.

Wallace, David (2012): The Emergent Multiverse: Quantum Theory According to the Everett Interpretation. Oxford: Oxford University Press.

Whewell, William (1847): The Philosophy of the Inductive Sciences, Volume II, Section III—Tests of Hypothesis. London: John W. Parker.

Whewell, William (1858/1968): Novum Organon Renovatum, being the Second part of the Philosophy of the Inductive Sciences, 3rd Edition, London. In Robert E. Butts (ed.) William Whewell's Theory of Scientific Method. Pittsburgh: University of Pittsburgh Press: 103-249.

Worrall, J. (1989): "Structural Realism: The Best of Both Worlds?" Dialectica (43): 99-124.

The two cultures—old and new debates on philosophy and the sciences

Brigitte Falkenburg

Institut für Philosophie und Politikwissenschaft, Technische Universität Dortmund, Emil-Figge-Straße 50, 44227 Dortmund, Germany

E-mail: brigitte.falkenburg@tu-dortmund.de

Abstract. The Neo-Kantian distinctions between science and the humanities (or cultural sciences), discussed around 1900, are instructive up to the present day. The philosophers then saw the distinguishing marks of the different sciences mainly in methodological aspects. The paper focuses on Windelband's distinction in terms of the nomothetic vs. idiographic method, Dilthey's criticism of it, and its further differentiations by Rickert and Max Weber. Rickert emphasised the significance of values in cultural science, whereas Weber bridged the methodological gap between the sciences in terms of ideal-typical explanations. The debates on the "two cultures" of the recent decades are still partially rooted in the old debates, and as far as they are not, striking similarities between the new and the old debates can be shown. Concerning the usefulness of the old debates for current philosophy of science, in particular Weber's approach sheds light on the role of idealizations and models in the sciences up to the present day.

1 Introduction

Philosophy of science has a long tradition as a meta-discipline that reflects on the conceptual foundations, methods, and contents of the sciences, as well as their significance for understanding nature and the place of human beings in nature. Considered a scientific discipline in its own right, philosophy belongs to the humanities, and so does its reflection on the sciences. In the late 19th century this reflection was subject to a philosophical debate between the Neo-Kantians and Dilthey that is instructive up to the present day, since it focuses on methodological aspects of the sciences. The starting point was Wilhelm Windelband's distinction between the exact sciences and the humanities in terms of the *nomothetic* and the *idiographic* method, its end point is marked by Max Weber's conception of *ideal-typical explanations*.[1] The scientific revolution in physics from 1900 onwards contributed to the decline of Neo-Kantianism and the rise of empiricist philosophy of science. 20th century history until the Second World War did the rest, and when the different cultures of the natural sciences and the humanities returned to the focus of philosophical debates in the 1960s, the Neo-Kantian tradition was forgotten.

---

[1]My sketch of the views of Cohen, Windelband, Rickert and Weber in this paper is based on Falkenburg 2020.

Science's Voice of Reflection, edited by G. Heinzmann & B. Löwe.
Comptes Rendus de l'Académie de Philosophie des Sciences I (2022), pp. 41–57.

In the following, I briefly sketch the background of that debate which arose in the 1890s when Dilthey sharply criticised Windelband's distinction between science and the humanities (§ 2). Rickert differentiated Windelband's distinction by emphasising the significance of values in the cultural sciences (§ 3). Finally, Max Weber bridged the methodological gap between natural and cultural science in terms of ideal-typical explanations (§ 4). The recent debates on the "two cultures" and the relation between philosophy and the sciences show continuities in content with the debates around 1900, even when there is no direct historical impact (§ 5). Finally, I attempt to give an outlook concerning the relevance of the Neo-Kantian approaches for current philosophy of science (§ 6).

2   The old debate: philosophy between naturalism and historicism

The distinction between the "two cultures", understood as different epistemic cultures of the exact sciences (or science) on the one hand and the human and/or cultural sciences (or humanities) on the other, came up in the late 19th century. The first half of the 19th century was marked by the opposition between post-Kantian German idealism from Fichte to Hegel and positivism in Comte's tradition, which spread with the progress of the natural sciences. The development of physics, chemistry, and biology lay the grounds for electrodynamics, thermodynamics, electrodynamics, atomism, and the theory of evolution. In parallel, the humanities underwent an enormous rise and led to the emergence of historicism as a counter-movement to positivism. In the second half of the 19th century, on the side of positivism, materialism and naturalism became influential, with Neo-Kantianism as a counter-movement. Later, the philosophy of life (*Lebensphilosophie*), in Germany represented in particular by Dilthey, added to these philosophical main streams.

The debate on science and the humanities which arose between Dilthey and Windelband in the 1890s resulted from the science wars of the late 19th century, one may say. These science wars had several fronts. They concerned the demarcations between philosophy, natural science, and the humanities; the opposition between naturalism and historicism; the status of psychology as well as social science between science and the humanities; and finally, the debates within social science (economics, law, and sociology) to which Weber contributed. Concerning the rise of psychology as a natural science, they were in particular fought as faculty disputes regarding the appointment of psychologists to philosophy chairs, around 1900 (Gundlach 2017). The case of sociology was to a certain extent similar to that of psychology, insofar as sociology was established at the universities as a new scientific discipline which employed mathematical methods, in the tradition of Comte's "social physics".

Dilthey and the Neo-Kantians had common grounds in their Kantian background and in their common opposition to positivism, materialism, and naturalism. Their efforts to demarcate the humanities against natural science were in particular connected with their attempts to counter (Dilthey) or combine (Windelband) the understanding of psychology as a natural science with an anti-naturalistic conception of the human mind. To the rise of psychology as a natural science, above all Helmholtz contributed with his research on the physiological foundations of human consciousness. He interpreted Kant's principle of causality as a structure of consciousness that is based on physiological processes and can be investigated by means of empirical psychology, based on experiments. The Neo-Kantians sharply criticized this program of naturalizing the human mind, and so did Dilthey.

Apart from these common grounds, the approaches of Dilthey and the Neo-Kantians substantially differed. Dilthey carried on Schleiermacher's foundation of hermeneutics and distinguished himself from Kant's transcendental philosophy by working out a historical account of reason, intended as a "critique of historical reason" opposed to Kant's *Critique of Pure Reason*. The Neo-Kantians of the Marburg and the Southwest schools carried on Kant's philosophy in different ways. In contrast to Dilthey, both schools relied on what Kant had called rational cognition, i.e., the program of giving foundations a priori to human cognition. A main difference between them was that Hermann Cohen, founder of the Marburg school, aimed at giving conceptual foundations to the exact sciences, whereas Windelband wanted to justify the principles of philosophy in terms of values. Being aware that the sciences and philosophy change with time, they could not avoid that their views about the rational foundations and the historical dimension of these disciplines were in a certain tension, which they attempted to resolve through opposing approaches.

Cohen (1914) tended towards constructivism and logicism in the sense of his logic of pure cognition (1902), even though he conceded that the "fact of science", in particular the well-established theories of physics, is subject to historical change. Cohen's constructivism was directed against the sensualistic conception of facts advanced by the empiricists of his day. Natorp (1910, 18) elaborated the genetic aspect of Cohen's constructivist approach, emphasising that scientific facts are not "given" (*gegeben*), but rather "posed as a task" (*aufgegeben*). Windelband (1882; 1883), on the other hand, marked a sharp distinction between contingent historical facts and universally valid values and focused on the validity of philosophical principles. In view of the many historical faces of philosophy, however, his approach tended towards a predominance of the historical elements of cognition. In a scale ranging from logicism (i.e., conceptualism concerning the foundations of theories) to historicism (i.e., empiricism concerning the

historical facts), Cohen was obviously closer to the extreme of logicism, or
constructivism, and Windelband closer to the extreme of historicism, or
empiricism. Neither of them convincingly succeeded in integrating the logical and the empirical aspects of science and philosophy in a comprehensive
account. Nor can they be blamed for this, considering that Kant did not
succeed either (as the philosophical debates on Kant's critical metaphysics
of nature show up to the present day).

Dilthey's historical account of human reason accented the rational and
the empirical elements of human cognition in a different way, claiming that
contingent historical facts can only be understood by subsuming them under
general concepts (Dilthey 1895/95). This claim was one of the reasons
for his sharp criticism of Windelband's distinction of natural science and
history, together with the fact that Windelband ranked psychology under
the natural sciences, taking into account the rise of experimental psychology
at the universities in a descriptive approach.

3   The nomothetic and the idiographic method

The debate began with Windelband's distinction between *nomothetic* and
idiographic disciplines (1894). Windelband explained his distinction between the "idiographic" and the "nomothetic" method of the empirical
sciences in his famous presidential address of 1894, *History and Natural
Science*. The distinction belongs to the philosophy of science, he considers
it "a theme from logic, especially from methodology, from the theory of
science" (Windelband 1894, 138).[2] For him logic is an applied discipline
employed in the practice of the sciences, which range from the "rational"
disciplines philosophy and mathematics to the "empirical sciences" (ibid.
141). Windelband emphasizes that the empirical sciences should not be
distinguished according to their objects 'nature' and 'mind', as the traditional distinction between natural science (Naturwissenschaft) and human
science (Geisteswissenschaft) indicates. Psychology in particular falls short
of this distinction, as a science that has the mind as its object but investigates it with the methods of natural science (ibid. 142). Instead, he suggests to classify the sciences according to their methods, distinguishing the
"nomothetic" from the "idiographic" method. The "idiographic" method is
historical and focuses on the description of individual facts. It comes close
to what he called the "genetic" method as opposed to the "critical" method
of philosophy (Windelband 1883). In contrast, the "nomothetic" method of
natural science is nomological, it aims at establishing general laws:

---

[2]My translation. The English translation of History and Natural Science by Guy
Oakes in Luft (2015, 287–298) is very instructive, but not sufficiently precise in detail
with regard to Windelband's philosophical terminology. Here and in the next quotation
I suggest a translation that is as literal as possible.

Here we now have before us a purely methodological classification
of the empirical sciences, which is to be based on reliable logical
terms. The principle of classification is the formal character of their
cognitive goals. Some of them search for general laws, others for
specific historical facts. [...].

Thus we may say: The empirical sciences search in the cognition of
reality either for the universal in the form of natural law or for the individual in the historically specified formation (*Gestalt*); in part they
consider the always invariable form; in part the individual, in itself
specified content of the actual events. The former are sciences of laws,
the latter are sciences of events; the former teach what always is, the
latter what once was. Scientific thinking is—if one is allowed to introduce new artificial expressions—in the one case *nomothetic*, in the
other *idiographic*. If we want to keep to the customary expressions,
we may further speak in this sense of the contrast between natural
sciences and historical disciplines. (Windelband 1894, 144–145; my
translation).

Windelband distinguishes between both methods according to their cognitive goals of capturing the logical subjects of universal judgments about
the general and invariable laws of nature, on the one hand, and singular
judgments about individual historical facts or events, on the other. He emphasizes that the respective distinction between natural sciences and historical disciplines is not strict, as the example of psychology shows, and that
there are scientific disciplines that combine both methods, in particular,
evolutionary biology.

A decade before Windelband's presidential address with this distinction, the first volume of Dilthey's *Introduction into the Human Sciences*
(1883) had appeared. Based on his own account of understanding in the
human sciences and on his way to developing hermeneutics as the appropriate method of understanding in the humanities, Dilthey strongly opposed
Windelband. He had three main objections: to the distinction as such,
given that there are natural sciences with idiographic elements and human
sciences with nomothetic goals; to the claim that psychology belongs to the
nomothetic disciplines; and finally, to the view that singular historical facts
may be understood as such, without embedding them in any general conceptual framework (Dilthey 1895/96). The first objection misses the approach,
given that Windelband himself admitted that the distinction is not sharp
and does not give rise to an unambiguous classification of the sciences. The
second objection makes a more substantial point, namely that Windelband,
with all his emphasis on the autonomy of historical methods, in relation
to psychology was not free from contemporary positivism (just like Cohen
in relation to mathematical physics as the predominant "fact of science").
But we may concede that his approach is descriptive, ranking psychology

among the natural sciences in face of the emergence of quantitative experimental psychology at the universities. From the point of view of philosophy of science, the third objection is much more substantial. Only Weber's later conception of ideal-typical explanations could counter it.[3]

Rickert presented a refined classification of the empirical sciences in *The Limits of Concept Formation in Natural Science* (1896/1902) and *Science and History* (1899), differentiating Windelband's distinction of the nomothetic and the idiographic method as follows. According to Rickert, the *subjects* of investigation should not be omitted in an adequate classification of the empirical sciences. Therefore, he adds them to Windelband's purely methodological distinction. With regard to the objects, he proposes to distinguish the natural sciences from the cultural (rather than human) sciences and to define the subjects of the cultural sciences in terms of values, following to a certain extent Windelband's distinction between the "critical" and the "genetic" method (Windelband 1883). He adopts Windelband's idiographic method with regard to the investigation of individual facts or events, calling it the "historical" method of. In this way, his approach results in distinguishing natural and cultural science in terms of the subjects of aswell as methods of investigation. Natural science refers to the phenomena of nature, and employs the nomothetic method to investigate them in search of universal laws of nature. Cultural science refers to values and employs the historical method to investigate their role in culture and society. Like Windelband, he stresses that these distinctions do not give rise to a sharp demarcation between the natural and the cultural sciences. On the one hand, many facts can be investigated from a nomothetic as well as a historical point of view. On the other hand, there are natural sciences such as evolutionary biology (Rickert 1896/1902, 280–282; 1899, 101–103) which proceed historically. For him, the empirical sciences are located in a continuum between the extremes of classical mechanics as the prototype of mathematical physics, and individual history as the prototype of a discipline focusing on facts only. These extremes meet in astronomy, which makes the individual celestial bodies subject to Newton's theory of universal gravitation (1896/1902, 285 and 444–448). In addition, he did not preclude that history also may become subject to relatively general laws, as far as the individual concepts applying to historical facts can be generalised (ibid. 490–492).

---

[3]For detailed accounts of the debate between Dilthey and the Neo-Kantians, cf. Makkreel and Luft 2010; Luft 2016; Makkreel 2021. For Dilthey's views about the natural sciences, cf. Pulte 2016 and Kühne-Bertram 2016. For Rickert and Dilthey, cf. Kinzel 2020; for Rickert and Weber, cf. Oaks 1990, Wagner and Härpfner 2015, Staiti 2018.

4   Ideal-typical explanations and the status of social science

For Rickert, the explanation of individual historical processes by general principles remained a vague possibility. Weber clarified this possibility. His conception of ideal-types aims at bridging the gap between historical descriptions and nomothetic explanations by means of a causal account of historical processes. In order to establish the objectivity of the social sciences, he considered social science as methodological hybrid between natural science and the humanities. On the one hand, he insisted that interpretive understanding is better suited to the subjects of the social sciences than capturing them with mathematical methods. One the other hand, he established the famous postulate of value neutrality, taking position against the normativity of sociology or economics.

Weber agreed with Rickert that values belong to the objects of cultural and social science, but that their scientific investigation should be value neutral. In addition, Weber agreed with Dilthey that historical facts and events cannot be understood without embedding them in an interpretative framework of regularities. In the extensive essay *"Objectivity" in Social Science and Social Policy* (Weber 1904), he starts to explain his conception of ideal-types and ideal-typical explanations for the economic concept of the market:

> We have in abstract economic theory an illustration of those synthetic constructs which have been designated as "ideas" of historical phenomena. It offers us an ideal picture of events on the commodity-market under conditions of a society organized on the principles of an exchange economy, free competition and rigorously rational conduct. This conceptual development brings together certain relationships and events of historical life into a complex, which is conceived as an internally consistent system. (89–90)

According to this explanation, the market as a historical phenomenon of which economics has an idealized typified conception, the ideal type. The ideal-type of the market is an "ideal picture", or model, of social actions under certain social conditions. This model is a "synthetic construct" of the dynamics of social life as an "internally consistent system" of social relations, i.e., an idealized model of the dynamics which occurs in a market under certain conditions, such as "the principles of an exchange economy, free competition and rigorously rational conduct". He continues:

> Substantively, this construct in itself is like a *utopia* which has been arrived at by the analytical (*gedankliche*) accentuation of certain elements of reality. Its relationship to the empirical data consists solely in the fact that where market-conditioned relationships of the type

> referred to by the abstract construct are discovered or suspected to
> exist in reality to some extent, we can make the *characteristic* fea-
> tures of this relationship pragmatically clear and *understandable* by
> reference to an *ideal-type*. (90)

The idealized model of the market is obtained by theoretical "accentua-
tion", picking out certain elements of reality and neglecting others, just as
the physicists do in their models of classical point mechanics or of an ideal
gas. The relation of such a model to empirical reality is the assumption that
the model relations, or "characteristic features" of the "ideal-type", refer "to
some extent" to the relationships which are "discovered or suspected to exist" in empirical reality. The model or "ideal-type" is a heuristic tool for
developing hypotheses:

> This procedure can be indispensable for heuristic as well as expository
> purposes. The ideal typical concept will help to develop our skill in
> interpretation in *research*: it is no 'hypothesis' but it offers guidance
> to the construction of hypotheses. It is not a *description* of reality but
> it aims to give unambiguous means of expression to such a description.
> (Ibid.)

Weber then passes from the example of exchange economy in modern
society to another case of an ideal-type, the economic model of a medieval
city, emphasising that such a model does not refer to average data but
only to certain ideal features of its object. The ideal type does not aim
at generating a statistical model of social phenomena. It takes up many
individual phenomena and condenses them into an ideal, abstract picture
of a cognitive object which as such does not exist in empirical reality. The
ideal-type is a "utopia", in the literal sense of something that exists nowhere:

> An ideal type is formed by the one-sided accentuation of one or more
> points of view and by the synthesis of a great many diffuse, discrete,
> more or less present and occasionally absent *concrete individual* phe-
> nomena, which are arranged according to those one-sidedly empha-
> sized viewpoints into a unified *analytical* construct (*Gedankenbild*).
> In its conceptual purity, this mental construct (*Gedankenbild*) cannot
> be found empirically anywhere in reality. It is a utopia. (Ibid.)

Finally, Weber emphasises the dynamic character of the relation between
model and empirical reality. The model has to be compared with the data,
i.e., the individual historical phenomena to which it refers. The comparison
works in two directions, going back and forth between the individual phenomena themselves and the idealized assumptions of the model. The model
helps to select the phenomena to which it applies, and the phenomena in
turn help to improve the model assumptions. In this way, the historical data

serve to modify the ideal-type in order to capture the structure of empirical reality more adequately (ibid.).

Furthermore, he adds a *causal* aspect to his conception of an ideal-type (Weber 1904, 93–110). To understand the historical dimension of social phenomena, he introduces genetic concepts which concern the predominant ideas, thoughts, or ideals of an epoch and their causal influence on the evolution of social phenomena, such as the church, the state, etc. An *ideal-typical explanation* then combines the ideal type of a historical constellation with a causal explanation. It reconstructs the causal process in which a specific social structure may have emerged under certain social conditions, such as the rise of capitalism in Western Europe under the condition of protestant ethics (Weber 1904-05).

Weber's ideal-typical explanations combine elements of Windelband's "nomothetic" and Rickert's "historical" method in the following way. An idealized model of a specific social phenomenon or historical constellation is constructed, which captures the *causally relevant factors* for the formation of that phenomenon or constellation. The model is *compared* with the empirical data from social reality, and it can be *modified* by adapting it to new data.[4]

5 The new debates: continuities and discontinuities

So far the old debates on science and the humanities and/or cultural sciences. In the 1920s, Neo-Kantianism lost importance. The philosophical debate on the sciences substantially shifted in face of the scientific revolutions of physics and the rise of logical empiricism. In Dilthey's tradition, Husserl's phenomenology became influential, upholding however the logical ideal of philosophy as a rigorous science. Cassirer transformed the approach of the Marburg school into his philosophy of symbolic forms, and Heidegger's Sein und Zeit appeared. The Davos dispute of 1929 between Cassirer and Heidegger, with Carnap in the audience, was as much an endpoint of the earlier philosophical debates as a milestone demarcating the diverging new traditions of 20th century philosophy.[5]

From 1933 onwards, the neo-Kantian tradition was completely cut off when Cassirer and other leading neo-Kantians lost their chairs and had to emigrate.[6] Only Weber continued to be discussed in sociology. In the philosophy of science, Carnap's logical empiricism dominated after the Second World War, competing with Popper's critical rationalism. The new debates

---

[4]For criticism of Weber's approach see Watkins 1952; for defence, Aronovitch2012 and Swedberg 2017.

[5]See Friedman 2000 and Gordon 2010.

[6]In particular, Richard Hönigswald lost his chair in Munich due to Heidegger's defamatory report (Schorcht 1990, 161).

on science, philosophy, and the humanities (or the cultural sciences) began in the 1960s. In our context, four of these debates are particularly relevant.

1. The positivism dispute between Adorno and Popper (Adorno et al. 1972) directly continued the old debates.[7] It began with Popper's talk and Adorno's comment at a conference of the German Sociological Association in 1961. The dispute concerned the methodology of sociology, took up the sociological debate on value neutrality to which Weber had contributed, and was carried out in the German-speaking world. Adorno did not distinguish between Popper's position of critical rationalism and the views of logical positivism and subsumed both under what Horkheimer (1937) called "traditional" theory, in contrast to the "critical" theory of the Frankfurt school. Horkheimer (1947) attacked the purely instrumental use of reason of the "traditional" theory, which he associated with Weber's account of purpose rationality and value neutrality.

2. The debate about the "two cultures" traces back to Snow's Lecture *The Two Cultures* (1959, 1963), which influenced the public discourse in the English-speaking world and far beyond. According to Snow, modern society is irremediably split into the cultures of literary intellectuals taking the attitude of backward-looking Luddites, on the one hand, and the scientists taking a forward-looking optimist attitude, convinced that any problem can be resolved by adequate means of science and technology, on the other hand. Snow took a crucial methodological point of the old debate up, without referring to it (and probably without knowing it, given that his background is the English intellectual culture, not the German philosophical tradition). Apart from the polemical connotations of labelling the scientific culture as progressive and the literary culture as regressive, it is justified to characterise the two cultures as future-oriented or backward-looking, respectively, insofar as science and technology aim at technical innovations and the humanities aim at understanding historical events, processes, theories, and works. This had precisely been the starting point for Dilthey's and Windelband's attempts to characterise the humanities as opposed to natural science.

Snow's lecture provoked polemical debate and his dictum of the "two cultures" began a life on its own, as can be seen from the German re-edition of Snow's essay (Kreuzer 1987), in a collection of articles with an enlarged scope of the discussion including the topic of science and responsibility. Today, strikingly many people are talking of the "two cultures" without having read a single line of Snow's many-faceted essay. This independence is mainly due to two further influential discussions that followed from the 1970s onwards and have very different topics. The related views

---

[7]Frisby (1972) discusses how the dispute relates to Weber's and Windelband's approaches.

have gained increasing influence on science policy in the last decades. One of them concerns science and responsibility, the other the relation between philosophy and history of science.

3. The discussion on science and responsibility had been opened after the Second World War, in view of the atomic bombs dropped on Hiroshima and Nagasaki. In the 1970s, the call for moral responsibility concerning science took new shape. The topics of the debate shifted towards the *Limits of Growth* (Club of Rome, 1972) and the environmental and related ethical problems posed by the use of science and technology. They were above all addressed in Jonas's *The Imperative of Responsibility* (1979). The very title of the book refers to Kant's categorical imperative. Indeed, when Jonas appeals to the responsibility of scientists in the face of the technical achievements made possible by their scientific work, he does not only emphasise that the problems created by technology have no technical solution. Moreover, he points out that ethical standards cannot be naturalised. In this respect, he is also in the tradition of Rickert, who saw the specificity of the cultural sciences in focusing on values.—The demand for responsibility resulted in claiming a new importance for the humanities as an ethical authority in a world dominated by science and technology. In German philosophy, Mittelstrass (1992) coined the dictum of the "Leonardo world" of science and technology, and he gave the call for responsibility a turn towards science policy, emphasizing the indispensability of the humanities for society. He distinguished between two complementary kinds of knowledge, the knowledge of disposal (*Verfügungswissen*), which serves the technical mastery of nature and is provided by science, and the knowledge of orientation (*Orientierungswissen*), which develops the guidelines for purposes and is owed to the humanities. These complementary kinds of knowledge refer to what there is and what ought to be, respectively. It obviously traces back to Kant's distinction between theoretical and practical reason. In addition, it recalls the neo-Kantian distinction between facts and values—and Weber's postulate of value neutrality which reminds us not to blur this distinction, for the sake of scientific objectivity.

4. Parallel to the debate on Snow's *Two Cultures*, Kuhn's book *The Structure of Scientific Revolutions* (1962, 1970) introduced history into the philosophy of science. Together with Feyerabend's *Against Method* (1975), it initiated the cultural turn of the philosophy of science and the humanities in general. Kuhn and Feyerabend challenged the sharp division between the two cultures by claiming that the development of science depends on human interests and social factors, just like any other human activity. Their work ushered in an era of anti-realism in the philosophy of science. Scientific realism was countered by social constructivism, scientific facts were seen as generated by the scientific community rather than as given in nature and

discovered by scientists. In this way social epistemology emerged, studying the external social factors of scientific practice and theory formation, and in particular the dependence of scientific practice on values in a given social context. The emergence of social epistemology led to a very special new clash of the two cultures, the new science wars (Sokal and Bricmont 1997; Carrier et al. 2004). This was certainly a fin de siècle phenomenon; in between, the debate has calmed down. Within philosophy of science, social epistemology has finally been established as a new field of research that investigates the impact of values on scientific research, meeting the field of science policy opened by the discussion about science and responsibility. Together with it, historical epistemology emerged focusing on the socio-cultural context of the evolution of the sciences since the end of the 19th century. Historical epistemology aims at reflecting the historical conditions under which scientific knowledge emerges by working through the traditions from the end of the 19th century to today's debates about the epistemic culture(s) of science (Rheinberger 2007).

Leaving the positivism dispute between Popper and Adorno aside, in the above accounts we encounter three completely different views of the relations between science and the humanities. According to Snow, science on the one hand and the humanities on the other constitute competing world views. According to Jonas and Mittelstrass, science and the humanities are complementary and both kinds of knowledge acquisition are indispensable for society. According to Kuhn, Feyerabend, and their followers, the distinctions between science and the humanities are blurred. We also encounter three different views about the role of philosophy as one of the humanities. In Snow's essay, philosophy is not present, or comes down to a general intellectual approach or attitude. For the followers of Kuhn and Feyerabend, philosophy of science amounts to social epistemology, complemented by historical epistemology. For Jonas and Mittelstrass, on the other hand, philosophy understood as ethics becomes the leading discipline of the humanities, whereas the function of the other humanities is not so clear. Needless to say, none of these approaches can capture the complex relationships between the natural sciences, humanities, and the social sciences in such a comprehensive a way as the works of the Neo-Kantians did, when taken together.

6 An outlook

The Neo-Kantian approaches to philosophy and the sciences in the Marburg and the Southwest School have one feature in common, notwithstanding all differences between Cohen, Natorp, and Cassirer on the one hand, Windelband, Rickert, and Weber (as far as we may consider him as a Neo-Kantian) on the other. All of them struggle with finding the balance between the

rational and the empirical or historical elements of scientific cognition, between general laws or principles and individual facts or events, between
Kant's principles *a priori* reflecting the structure of Newtonian science and
the historical stage of the empirical sciences.

From today's perspective, this conflict has not been resolved, but rather
intensified, insofar as today's natural sciences have to deal more than ever
with the theory-ladenness of empirical data. This is particularly true in
the age of big data, in view of the methods of machine learning employed,
e.g., in the data analysis of the experiments and measurements of particle
physics, astrophysics, or astroparticle physics. Here, to a certain extent the
considerations of the Neo-Kantians from the Marburg school from Cohen to
Cassirer come to bear, as a version of constructivism that does more justice
to the methods of the sciences than recent social constructivism did.

Concerning the humanities, Windelband's account of the "idiographic"
method has long been abandoned in favour of Dilthey's distinction between
explaining and understanding, which Weber took at least partially into account in his conception of ideal-typical explanations. Indeed, Dilthey's distinction is influential in the humanities up to the present days, not least
thanks to von Wright's *Explanation and Understanding* (1971).

Windelband's account of the "nomothetic" method is obviously close to
the later deductive-nomological (DN) model of explanation (Hempel 1965),
which is however much more precise. But the DN model of explanation,
notwithstanding its elaborations (Woodward and Ross 2021), was challenged in more recent philosophy of science in two regards. On the one
hand, Cartwright (1983) argued that even the laws of physics from Newton's theory of gravitation to quantum mechanics lack universality. On the
other hand, Morrison and Morgan (1999) showed that modelling in physics
has much more in common with the models of economics than usually ac-
knowledged.

Indeed, Weber's conception of ideal-typical explanations is very close to
this approach. Weber's ideal types are indeed models as mediators in the
sense of recent philosophy of science. They make it possible to go back
and forth between data and theories in order to develop more differentiated
models. His ideal-typical explanations aim at an idealized reconstruction
of historical phenomena and the way in which they arise, rather than at a
naturalization of social phenomena in terms of the statistical behaviour of
social agents. His conception of ideal-typical explanations anticipates the
insight of recent philosophy of science that models are instruments to investigate empirical reality rather than giving true descriptions of it (Morgan
and Morrison 1999). Models mediate between the phenomena and abstract
theories, making it possible to go back and forth between the poles of the
phenomena (empirical data) and rational cognition (theory), in order to im-

prove the models and their theoretical foundations. To my view, Weber's ideal-typical explanations deserve much more attention in current philosophy of science and the humanities.

References

Adorno, Theodor W., et al. (eds.) (1972): Der Positivismusstreit in der deutschen Soziologie. Darmstadt: Luchterhand.

Aronovitch, Hilliard (2012): Interpreting Weber's Ideal-Types. In: Philosophy of the Social Sciences 42, 356–369.

Cartwright, Nancy (1983): How the Laws of Physics Lie. Oxford:Clarendon Press.

Cohen, Hermann (1914): Das Verhältnis der Logik zur Physik. In: Einleitung mit kritischem Nachtrag zur 9. Auflage der Geschichte des Materialismus von Friedrich Albert Lange. Leipzig: Brandstetter, 58–94. Repr. in: Werke, Band 5.II. Hildesheim: Olms 1977. Quoted after: The Relation of Logic to Physics from the Introduction, with Critical Remarks, to the Ninth Edition of Lange's "History of Materialism". Engl. transl. by Lydia Patton, in: Luft (2015), 117–136.

Cohen, Hermann (1902): System der Philosophie. Erster Teil: Logik der reinen Erkenntnis. Berlin: Cassirer. Nachdruck d. 2., verb. Aufl. (1914) in: Werke, Band 6. Hildesheim: Olms 1977.

Damböck, Christian, and Hans-Ulrich Lessing (eds.) (2016), Dilthey als Wissenschaftsphilosoph. Freiburg: Alber.

Dilthey, Wilhelm (1883): Einleitung in die Geisteswissenschaften. Versuch einer Grundlegung für das Studium der Gesellschaft und der Geschichte. Bd. 1. Leipzig: Duncker & Humblot.

Dilthey, Wilhelm (1895/96): Über vergleichende Psychologie / Beiträge zum Studium der Individualität. Repr. in: Gesammelte Schriften Vol. V, Göttingen: Vandenhoeck & Ruprecht 1924, 241–316. Engl. transl.: Contributions to the Study of Individuality. In: R.A. Makkreel and F. Rodi (eds.), Selected Works Vol. II, Princeton, NJ: Princeton University Press, 2010, 211–284.

Falkenburg, Brigitte (2020): On Method: The Fact of Science and the Distinction between Natural Science and the Humanities. In: D. Heidemann (ed.): Kant Yearbook 2020: Kant and Neo-Kantianism. Berlin: de Gruyter, 1–31.

Friedman, Michael (2000): A Parting of the Ways. Carnap, Cassirer, and Heidegger. Chicago: Open Court.

Frisby, David (1972): The Popper-Adorno Controversy. The Methodological Dispute in German Sociology. In: Philosophy of the Social Sciences 2, 105–119.

Gordon, Peter E. (2010): Continental Divide: Heidegger, Cassirer, Davos. Harvard University Press.

Gundlach, Horst (2017): Wilhelm Windelband und die Psychologie. Das Fach Philosophie und die Wissenschaft Psychologie im Deutschen Kaiserreich. Heidelberg: University Publishing.

Hempel, Carl G. (1965): Aspects of Scientific Explanation and Other Essays in the Philosophy of Science, New York: Free Press.

Horkheimer, Max (1937): Traditional and Critical Theory. Repr. in: Critical Theory: Selected Essays, New York: Herder & Herder 1972, 188–243.

Jonas, Hans (1979): Das Prinzip Verantwortung. Versuch einer Ethik für die technologische Zivilisation. Frankfurt am Main: Insel. Engl. transl.: The Imperative of Responsibility: In Search of Ethics for the Technological Age. Chicago: University of Chicago Press 1984.

Kinzel, Katherina (2020): Neo-Kantianism as hermeneutics? Heinrich Rickert on psychology, historical method, and understanding. In: British Journal for the History of Philosophy, 29, 614–632.

Kuhn, Thomas S. (1962): The Structure of Scientific Revolutuions. 2nd ed. Chicago: Chicago University Press 1970.

Kühne-Bertram, Gudrun (2016): Zum Verhältnis von Naturwissenschaften und Geisteswissenschaften in der Philosophie Wilhelm Diltheys. In: Damböck and Lessing (2016), 225–248.

Luft, Sebastian (ed.) (2015): The Neo-Kantian Reader. London and New York: Routledge.

Luft, Sebastian (2016): Diltheys Kritik an der Wissenschaftstheorie der Neukantianer und die Konsequenzen für seine Theorie der Geisteswissenschaften. Das Problem des Historismus In: Damböck and Lessing (2016), 176–198.

Makkreel, Rudolf A. (2010): Wilhelm Dilthey and the Neo-Kantians: On the Conceptual Distinction between Geisteswissenschaften and Kulturwissenschaften. In: Rudolf A. Makkreel and Sebastian Luft (eds.), Neo-Kantianism in Contemporary Philosophy. Bloomington: Indiana University Press, 253–271.

Makkreel, Rudolf A. (2021): "Wilhelm Dilthey", The Stanford Encyclopedia of Philosophy (Spring 2021 Edition), Edward N. Zalta (ed.).

Makkreel, Rudolf A., and Sebastian Luft (2010): Dilthey and the Neo-Kantians: The Dispute over the Status of the Human and Cultural Scienses. In: D. Moyar (ed.), The Routledge Companion to Nineteenth Century Philosophy. London: Taylor & Francis Group, 554–597.

Mittelstrass, Jürgen (1992): Leonardo-Welt. Über Wissenschaft, Forschung und Verantwortung. Frankfurt am Main: Suhrkamp.

Morgan, Mary S., and Margaret Morrison (1999): Models as Mediators: Perspectives on Natural and Social Science. Cambridge University Press.

Natorp, Paul (1910): Logische Grundlagen der exakten Wissenschaften. 2nd ed.: Leipzig und Berlin: Teubner 1921.

Oakes, Guy (1990): Weber and Rickert: Concept Formation in the Cultural Sciences. Cambridge, Mass.: MIT Press. German ed.: Die Grenzen kulturwissenschaftlicher Begriffsbildung : Heidelberger Max-Weber-Vorlesungen 1982. Frankfurt am Main: Suhrkamp 1990.

Pulte, Helmut (2016): Gegen die Naturalisierung des Humanen. Wilhelm Dilthey im Kontext und als Theoretiker der Naturwissenschaften seiner Zeit. In: Dambök and Lessing (2016), 63–85.

Rheinberger, Hans Jörg (2007): Historische Epistemologie zur Einführung. Dresden: Junius. Engl. transl.: On Historicizing Epistemology. An Essay. Stanford: Stanford University Press 2010.

Rickert, Heinrich (1896/1902): Die Grenzen der naturwissenschaftlichen Begriffsbildung. Eine logische Einleitung in die historischen Wissenschaften. Part I: 1896, Part II: 1902. 6th improved edition. Tübingen: Mohr Siebeck 1929. Abridged Engl. transl.: The Limits of Concept Formation in Natural Science, Cambridge University Press 1986.

Rickert, Heinrich (1899): Kulturwissenschaft und Naturwissenschaft. 6th and 7th expanded editions, Tübingen: Mohr Siebeck 1926. Engl. transl.: Science and history: A critique of positivist epistemology, Princeton: Van Nostrand 1962.

Schorcht, Claudia (1990): Philosophie an den Bayerischen Universitäten 1933–1945. Erlangen: Harald Fischer.

Sokal, Alan, and Jean Bricmont (1997): Impostures Intellectuelles. Paris: Jacob.

Snow, Charles P. (1961): The Two Cultures and the Scientific Revolution. The Rede Lecture. 1959. New York: Cambridge University Press.

Staiti, Andrea (2018): "Heinrich Rickert", The Stanford Encyclopedia of Philosophy (Winter 2018 Edition), Edward N. Zalta (ed.).

Swedberg, Richard (2017): How to use Max Weber's ideal type in sociological analysis. In: Journal of Classical Sociology 18, 181–196.

von Wright, Georg Henrik (19721): Explanation and Understanding. Ithaca: Cornell University Press.

Wagner, Gerhard, and Claudius Höpfner (2015): Neo-Kantianism and the social sciences: from Rickert to Weber. In: Nicolas de Warren and Andrea Staiti (eds.), New Approaches to Neo-Kantianism. Cambridge University Press.

Watkins, J. W. N. (1952): Ideal Types and Historical Explanation. In: The British Journal for the Philosophy of Science 3, 22–43.

Weber, Max (1904): Die „Objektivität" sozialwissenschaftlicher und sozialpolitischer Erkenntnis. In: Johannes Winckelmann (ed.), Max Weber: Gesammelte Aufsätze zur Wissenschaftslehre. 3rd, extended ed. Tübingen: Mohr Siebeck 1968, 146–214. Quoted after the Engl. transl.: "Objectivity" in Social Science and Social Policy. in: E. A. Shils and H. A. Finch (eds.), Max Weber: The Methodology of the Social Sciences. New York: Free Press, 1949, 50–112.

Weber, Max (1904-05): Die protestantische Ethik und der „Geist" des Kapitalismus. In: Archiv für Sozialwissenschaft und Sozialpolitik 20 (1904), 1–54, and 21 (1905), 1–110. Engl. Transl. by T. Parsons: The Protestant Ethic and the Spirit of Capitalism. London: Routledge 1992.

Windelband, Wilhelm (1883): Kritische oder genetische Methode? In: Windelband (1915), Vol. II, 99–135. Quoted after: Critical or Genetic Method?
Engl. Transl. by Alan Duncan, in: Luft (2015, 271–286).

Windelband, Wilhelm (1894): Geschichte und Naturwissenschaft. (Straßburger
Rektoratsrede) In: Windelband (1915), Vol. II, 136–160. Quotations: my
translation. Engl. transl. by Guy Oakes: "History and Natural Science"
(Presidential Address Strasbourg), in: Luft (2015, 287–298).

Windelband, Wilhelm (1915): Präludien. Aufsätze und Reden zur Philosophie
und ihrer Geschichte. Fünfte, erweiterte Auflage (two Vols.). Tübingen:
Mohr Siebeck.

Woodward, James and Lauren Ross (2021): Scientific Explanation. In: Edward
N. Zalta (ed.). The Stanford Encyclopedia of Philosophy (Summer 2021
Edition).

Engaging or not engaging with transdisciplinary research: on methodological choices in philosophical case studies

Inkeri Koskinen

Philosophy, Faculty of Social Sciences, Tampere University, Kanslerinrinne 1, 33014 Tampere, Finland

E-mail: `inkeri.koskinen@tuni.fi`

Abstract. Conducting empirical case studies in philosophy of science entails methodological decisions—decisions that can limit the ways in which philosophers can engage with and have an impact on the science they are studying. In this paper I approach such limitations through two examples: case studies in which philosophers of science used qualitative methods in the study of inter- and transdisciplinarity.

1 Introduction

Conducting an empirical case study is a singular way for a philosopher of science to engage with scientists. Here I am naturally talking about the kind of case studies that can include actual engagement: participant observation, interviews, perhaps even co-research in collaboration with scientists. In short, philosophical work that adopts methods from the social sciences, but uses them for the purposes and interests of philosophy of science. Such case studies have become fairly common in philosophy of science, as approaches such as philosophy of science in practice (Ankeny et al. 2011; Boumans & Leonelli 2013; Chao & Reiss 2017) or empirical philosophy of science (Wagenknecht, Nersessian & Andersen 2015) have become popular in the field. When a philosopher spends months or perhaps even years getting familiar with the work of a research group, a project, or for instance the work conducted at some research institute, they get a rare opportunity to build connections with the scientists they are studying. And as cultivating interpersonal interactions with scientists appears to be a particularly efficient way for philosophers to have an impact on science (Plaisance et al. 2021), one would think that conducting empirical case studies would be a straightforward way to take the role of the "voice of reflection" within the group or community they are studying—as noted in the description of the AIPS 2018 conference—raising questions about the "motivation, norms, values, methods, and limitations of the scientific enterprise".

However, the issue is not as simple as that. The aims of philosophy of science are typically normative, and taking the role of the voice of reflection should, if successful, result in changes in the science that is being studied. Such an outcome may be hard to reconcile with the epistemic aims of a case study. If a philosopher wants to learn something that can be studied by conducting a case study, having a significant impact on the science one is studying can sometimes be counterproductive. Adopting methods from the social sciences entails the need to make make methodological choices, and these choices can involve both epistemic and research ethical considerations that lead to the conclusion that one should not attempt to influence the science one is studying.

Methodological choices and decisions can thus restrict the ability of a philosopher of science to have an impact on the science they are studying. But what kind of restrictions are we talking about here? In this paper I approach this question through two examples: case studies in which philosophers of science used qualitative methods in the study of inter- and transdisciplinarity. One of these case studies I conducted on my own, and in the other I was a member of an interdisciplinary research team that included philosophers and STS scholars. I will argue that the normative aims of philosophy of science should very much influence the methodological decisions made when conducting case studies. We cannot simply adopt empirical methods from the social sciences; we must critically examine them in order to understand how they will restrict our work as philosophers of science, and decide whether the limitations are worth it, and continue developing the methods we decide to use.

I will begin by briefly examining the use of empirical case study methods in contemporary philosophy of science. I then introduce the broad topic of the two case studies I will be discussing: transdisciplinarity. After describing the case studies, I will conclude by considering the advantages and limitations of the methodological choices made in these two cases.

2   Engaging with science while studying it

In a recent paper based on interviews with 35 philosophers of science, Kathryn S. Plaisance, Jay Michaud and John McLevey (2021) come to the conclusion that face-to-face or interpersonal interactions are the most important pathway through which philosophy of science has an impact on science, science policy, or science education. This emphasis on the importance of active engagement is further reinforced, for instance, in the recent book on "field philosophy" edited by Evelyn Brister and Robert Frodeman. Direct involvement and interventions are effective. Doing philosophical work "that is directly engaged in problem-solving and that explicitly demonstrates its real-world effects" (Brister & Frodeman 2020, 2; see also Plaisance & Elliott 2021) is an efficient way for philosophers to bring about change. In philosophy of science, this means engagement with the scientific endeavour.

Such engagement can take many forms. Philosophers can, for instance, collaborate with scientists (see, e.g., Keven et al. 2018; Bursten 2020; Beck et al. 2021) or take part in the development of science policy initiatives (see, e.g., Vermier et al. 2018; Parker & Lusk 2019)—or they can study the ways in which scientists work, often using and further developing methods adopted from other fields (e.g., Wagenknecht, Nersessian & Andersen 2015; Robinson, Gonnerman & O'Rourke 2019). The multiplicity of forms of engagement reflects the multiple ways in which philosophers wish to impact science. Plaisance, Michaud and McLevey (2021) identify six central types of impact philosophy of science can have: analyzing concepts or issues in a scientific field; identifying problems with scientific methods, inferences, and explanations, and offering alternatives for scientists to consider; highlighting the role of values in science; contributing to the development of new scientific knowledge; enhancing science policy and legislation; and improving science education. These types of impact reflect the thorough normativity of our field. We try to influence science so that it would better reflect the epistemic, ethical, and social ideals we believe it should manifest. As Angela Potochnik (2018) sums up, even when doing strongly engaged, practice-based philosophy of science, and even when taking seriously "what scientists actually do, using these practices as the starting points for our philosophical accounts of the aims, processes, and products of science", we must not be shy of arguing against scientists: "philosophers of science not only can but indeed *must* bring to bear considerations that go beyond existing scientific practices". This kind of normativity is crucial if philosophy of science is to actually influence science.

The emphasis on taking seriously what scientists do, and paying attention to scientific practices, results from the naturalistic and social turns in the philosophy of science. Many philosophers of science today argue that it is necessary to analyse real scientific practices before presenting philosophical claims or theories, or normative considerations about science. With the growing importance of the social epistemology of scientific knowledge, and more recently, the realisation that not only social practices, but also institutional configurations shape science and scientific knowledge, naturalistically oriented philosophers of science have started paying attention not only to the work of individual scientists, but also to the social and institutional aspects of the scientific endeavour. Doing philosophical work on them requires not only engagement, but also research—"conscious, detailed, and systematic study of scientific practice that nevertheless does not dispense with concerns about truth and rationality" (Ankeny et al. 2011, 304). Often such "detailed and systematic study" means introducing qualitative methods, adopted from various social sciences, into philosophy of science (Kosolosky 2021; Boumans & Leonelli 2013; Wagenknecht, Nersessian & Andersen 2015). Collaborations with people trained in the use of such methods—for instance, sociologists of science or STS scholars—are also relatively common by now.

The use of qualitative methods does not preclude active engagement that attempts to influence the science that is being studied. Some qualitative methods—even case study methods—used in the social sciences allow engagement and participation. A philosopher could, for instance, end up doing some kind of co-research with scientists, or participate in a transdisciplinary project, thus taking part in the scientific endeavour they want to understand (for more on such methods, see, e.g., Hartley & Benington 2000; Hirsch Hadorn & al. 2008; Schrögel & Kolleck 2019). Such engagement can offer valuable opportunities for influencing the development of scientific practices, or for instance science policy. However, some questions are best studied through case studies where researchers do not attempt to influence the processes they are studying, but actively avoid doing so. When conducting a case study, a philosopher of science does not necessarily wish to have a significant impact on its results.

Adopting qualitative methods from the social sciences means that philosophers of science have to make new kinds of methodological choices and decisions. As yet, the methodological work in practice-oriented, "empirical" philosophy of science is in its early stages. Here I attempt to contribute to the development of these methods by examining the relationship and possible tensions between the normative aims of philosophy of science, and some methodological and research-ethical considerations that suggest caution with regard to influencing one's object of study. While case study methods offer useful tools when philosophers of science wish to influence science and science policy, they also impose limitations. I will now explore such limitations by discussing two case studies where philosophers of science had a clear opportunity to influence evolving practices or institutional changes, but to different extents refrained from doing so.

3   Transdisciplinary research

In both of the cases I will be discussing in the remainder of this paper, my focus was on research that could be called transdisciplinary. The term "transdisciplinarity" has many partially overlapping meanings. Moreover, transdisciplinary research shares many characteristics with other approaches that stress societal impact and stakeholder engagement, and quite often a project that is called citizen science or co-research could also be called transdisciplinary. However, one of the central developers of transdisciplinarity, Christian Pohl (2011), identifies four features that are central to the approach: the search for a unity of knowledge, a focus on socially relevant issues, transcending and integrating disciplinary paradigms, and the inclusion of extra-academic partners in the research process. Building on systems theory and the "Mode-2" concept of knowledge production, transdisciplinary research emphasises the "integration, assimilation, incorporation, unification and harmony of disciplines, views and approaches" (Choi & Pak 2006, 356).

Uskali Mäki and I have drawn together these and some other available definitions (see, e.g., Pohl & Hirsch Hadorn 2007; Leavy 2011), and compiled a list of attributes that are often mentioned when characterising transdisciplinary research. Transdisciplinary research, then, is research that transcends scientific disciplines and/or approaches within academia, integrates academic disciplines and/or approaches with one another, addresses and attempts to solve socially and practically relevant issues, involves extra-academic agents in various roles, involves and integrates academic and extra-academic knowledges, values, and interests, and serves "the common good" or some similar goal (Koskinen & Mäki 2016, 424). According to its advocates, transdisciplinarity is needed because the adequate understanding and solving of many pressing, complex problems—often simultaneously environmental and social ones—requires the integration of diverse perspectives, knowledges, and skills. (Zierhofer & Burger 2007; Hirsch Hadorn et al. 2008; Brown et al. 2010; Carew & Wickson 2010; Hirsch Hadorn, Pohl & Bammer, G. 2010; Adler et al. 2018; Koskinen & Rolin 2022.)

Both inter- and transdisciplinarity are methodologically ambitious, as their aim is often stated to be the integration of different approaches, methods and perspectives. In transdisciplinary research, this involves not only scientific perspectives, but the viewpoints of the extra-academic participants are also "included in the first stage of problem framing, ensuring that the questions addressed by research will be relevant, i.e. salient, and results credible, i.e. evidence appropriate for the particular policy problem" (Adler et al. 2018, 184).

To summarise, transdisciplinarity is solution-oriented research where the problems are framed in cross-disciplinary and even extra-academic terms, and researchers from many fields are involved in the search for solutions, often also with extra-academic partners. In contemporary science policy, inter- and transdisciplinarity are often taken to be efficient and sorely needed ways to approach and solve pressing societal and environmental problems (Maassen & Weingart 2005; Maassen & Lieven 2006; Jacobs & Frickel 2009; Huutoniemi et al. 2009; Pohl, Truffer & Hirsch Hadorn 2017). This belief has, in many countries worldwide, led to institutional and organisational changes that are ment to encourage and incentivise scientists towards inter- and transdisciplinary collaborations.

If a philosopher of science is to study transdisciplinarity, and wishes to discuss some actual examples, historians or even sociologists of science have relatively little to offer. The approach is so new that the existing studies describing and analysing it do not well lend themselves to the use of a philosopher of science. Therefore, conducting a philosophical case study is tempting. And that is what I ended up doing. I conducted one on my own, and took part in a larger one. I followed, from the beginning to the end, a two-year project that involved social scientists from several fields, journalists and artists. And as a member of an interdisciplinary team involving philosophers and STS scholars, I participated in a study of a technical university where research was being reorganised into strategically designed inter- and transdisciplinary research platforms. In the first case study—let us call it SocJournArt—the project I studied focused on social inequality in Finland. As I will describe in more detail below, the social scientists collaborated closely with the journalists and the artists, particularly in data collection. My initial aim was to get, through this one example, a better understanding of how questions of demarcation and decisions about the epistemically relevant criteria used in knowledge production can be negotiated in a research team that has extra-academic experts in important roles.

In the second and much larger case study—which we named BizTech—philosophers and STS scholars joined forces in order to study the structural reorganisation of research at a small technical university in a Nordic country. We were particularly interested in diverse tensions that arise from such an institutional change, and from the shift from discipline-driven to more demand-driven university research.

In both projects it was soon clear that we had to decide what kind of input philosophers could and should offer to the researchers and organisations being studied. In BizTech we studied several research platforms, and in addition to tensions arising in inter-and transdisciplinary collaborations, we were interested in the changing institutional context where the inter- and transdisciplinary knowledge production happened. Our work was seen as potentially relevant to science policy, and both some members of the university's upper management and major sources of research funding were interested in our results, even preliminary ones. In SocJournArt I concentrated on just one project, and as I was allowed to participate in project meetings and online discussions, the participants would not only be interested in what I was doing in their project, but at times they also wanted to tap into my expertise—after all, in transdisciplinary research, all participants are generally supposed to take part in the development of a shared understanding of the problem at hand. It soon became clear to me that I was expected to participate—remaining silent would not do.

In both cases there were methodological and/or research-ethical reasons to refrain from offering any comments. In both cases there were also good reasons for disregarding some of these reasons, and for offering views and informed opinions. The decisions we made in BizTech were different from the ones I made in SocJournArt. I will now describe the cases and our choices in more detail.

4   BizTech: keeping the distance

In the project *Interdisciplining the university—Prospects for sustainable knowledge production* (2016–2021), led by Mikko Salmela, our research team conducted a large case study at a small technical university ("BizTech") which was undergoing a significant structural reorganisation. During the study period, BizTech implemented a university-wide policy that was meant to incentivise inter- and transdisciplinary collaboration, and to make the university more competitive in the pursuit of EU research funding. All internal research funds were reallocated to temporary research platforms that had to incorporate researchers from at least two of the university's three schools. Collaboration with diverse stakeholders was also encouraged.

Our team included philosophers—mostly philosophers of science—and STS scholars, and our aim was to explore how the reorganization of research into strategically designed inter- and transdisciplinary research platforms would influence the dynamics of knowledge production. In our project we were particularly interested in the diverse tensions—including epistemic, structural and emotional ones—that arise when such a change is implemented (see Mansilla et al. 2016; Parker & Crona 2012; Turner et al. 2015; Salmela & Mäki 2018), and their epistemically significant repercussions. The overarching aim was to evaluate the consequences of such a structural reorganisation for the epistemic sustainability of university research. Is it possible to push for demand-driven, solution-oriented inter- and transdisciplinarity through internal funding in an epistemically sustainable way? The question is relevant for science policy, because many European universities are currently redirecting their internal research funding in similar ways, as university administrators hope that this will increase their chances in getting Horizon Europe funding (Salmela, MacLeod & Munck af Rosenschöld 2021; see also Lindvig & Hillersdal 2019).

Our group followed the development of the platforms from 2015 onward, conducting semi-structured interviews ($n \approx 50$) with platform principal investigators, professors, coordinators, and researchers from three platforms, and the university management. The last interviews were finished in the spring 2021. The analysis of this data, and of the other data our team collected (e.g., documents such as research plans and evaluation reports) is still ongoing.

Already when we started planning the project, it was clear that we had to make some important decisions: how much and to whom would we talk about our work during the project? In BizTech, some members of the university management were naturally interested in our findings. And on the level of national science policy, the reorganisation in BizTech was seen as an organisational experiment, the results of which we were studying. In other words, we had a chance of influencing both the organisational restructuring we were studying, and possibly also science policy. On the other hand, there were very clear methodological and research ethical reasons for being careful about the information we would disclose.

Firstly, we collected legally confidential material, such as research plans and evaluation reports, access to which required an official permission from the BizTech administration. Our research team therefore signed an agreement concerning our access to and use of this material.

Secondly, during the interviews, we wanted to ask our informants about the platforms in which they were involved, and wanted to learn about their experiences and even their feelings regarding the work at the platforms. We would hardly have received the kind of answers we were after if it would have seemed that we were reporting to the university management. We therefore made it very clear right from the start that the BizTech management had no control over our problem setting or the selection and analysis of our data, and they had no special access to or control over our findings or possible recommendations.

Thirdly, the anonymity of our interviewees and other participants had to be protected, and any material that was privileged by nature, such as unpublished results or personal discussions, had to be kept private. Such anonymity was particularly important because we were studying an environment where we were already expecting tensions to arise, and any carelessness on our side could have intensified suspicions or envy within the research community. For these reasons we were throughout the project quite cautious when talking about it in public, and did not seize all the available opportunities for attempting to influence science policy. Now that the project has ended and we have begun publishing our results, we have also started presenting our findings in science policy arenas. Among our most important results is the observation that the kind of 'strategically incentivised organisational interdisciplinarity' (as we have ended up calling it) we studied does not always have particularly much in common with interdisciplinarity as it is described in handbooks and science policy briefs (see Salmela, MacLeod & Munck af Rosenschöld 2021).

5   Social scientists, journalists, and artists: participating, cautiously

The other case study I conducted on my own. I followed a two-year (2015–2017) research project, SocJournArt, where social scientists collaborated with journalists and artists. I started following it already during the application process, which begun when a foundation called for research projects that would study social justice and inequality in Finland. Collaboration with journalists was demanded in the call, and the foundation in question also favours collaborations between scientists and artists. The team designed a project that would continue and expand on already established collaborations between some of the social scientists and journalists, and would also include photographic artists. When I learned about the plan, I asked if I could follow the project, as I was very interested particularly in seeing how the collaboration between the social scientists and the artists would work out, as it was something quite new to everyone involved. How would the team members come up with the shared principles and criteria they would need in order to collaborate? How would they reach a shared understanding of their research topic, social inequality in Finland? Would they?

Once the project got funding, I participated in research meetings, followed the group's lively online discussions, read the project publications—both academic and journalistic—and participated in the closing workshop, where I interviewed everyone who was present. Later, once the final art exhibition had opened, I conducted some complementary interviews. I focused on the two sub-projects that included collaboration with journalists and artists, and conducted a total of eight semi-structured interviews with everyone who had taken part in one or both of these sub-projects, as well as with some of the other members of the research team.

In both of the two sub-projects I followed, social scientists collaborated with the extra-academic participants in data collection. The journalists and the social scientists conducted a large survey in a major newspaper. They jointly designed the survey, building on their previous experiences of collaboration, and the journalists wrote several articles about the results. The artists led a one-year artistic workshop with both lay participants and professional photographers. In addition to the artistic work, this group was supposed to produce data for visual sociology. The artists who led the workshop collaborated with the social scientists in designing and organising the data collection.

Quite early on I realised that I could not remain a passive observer. Particularly the collaboration between the social scientists and the artists had to start from scratch: the participants had little to no previous experience of such collaborations, and they welcomed and even demanded the contributions of a philosopher of science familiar with the multifaceted literature on research collaborations across the boundaries of science. I was there because I wanted to observe how social scientists of the quantitative ilk manage to collaborate with artists. But soon the emphasis in my participant observation started to be more on the side of "participant" than I had originally envisioned. On the one hand, I was somewhat worried about distorting the data I wished to gather. On the other, the participants wanted me to contribute, and I felt that as a philosopher of science, I should respond to such a demand, and offer ideas and arguments. Whereas in BizTech our team was in a position where we might have been able to influence organisational practices at the university we studied, in SocJournArt I could influence study design, and facilitate the collaboration I wanted to study.

In the end I decided to participate, but cautiously. When I was asked for feedback and suggestions, I would point out ideas and options that were readily available in the literature on transdisciplinarity, co-research, participatory research, and other forms of research collaborations between scientists and extra-academic experts. I felt this to be useful, as many of the participants were not well versed in that literature, and I thus saved them some time, but did not affect the outcome in too significant ways. Following the project also gave me some interactional expertise—I was in a position where I could sometimes offer useful comments. And because of the transdisciplinary nature of the collaboration, the non-existence of extablished collaborative practices between the participants, and the need to build a shared framework for the project, the participants were willing to hear me.

In the end I was lucky. I formed my most significant critical arguments regarding the project only after it had already ended. It was during the interviews I conducted at the closing workshop of the project that I finally started to understand the most important bone of contention and source of confusion between the social scientists and the artists. To describe it briefly: for the social scientists, a photograph was evidence of the thing pictured. For the artists, a photograph was evidence of the choices made by the person who took the picture. This disagreement, which for some time remained unclear for everyone involved, resulted in disagreements about how to plan the data collection. This in turn delayed the data collection so much that the project ended before the gathered data could be analysed. But in the end, for the participants of SocJournArt this mattered much less than I would have anticipated. While the collaborative data collection stagnated, the artistic workshop did impressive work on its own, and the project greatly benefited from the public attention that the final art exhibition received. (Koskinen 2018a; Koskinen 2018b; Koskinen under review.)

6   Coordinating methodological decisions with normative aims

Conducting empirical research in philosophy of science entails methodological decisions. And these decisions limit the ways in which philosophers can engage with and have an impact on the science they are studying. In SocJournArt I could have adopted collaborative methods and participated in the project fully (for an ambitious example, see Ginsberg et al. 2014). But had I concentrated more on facilitating the collaboration between the artists and the social scientists, I most likely would not have realised that the members of the research team cared much more about the societal impact they were creating—regardless of how it was created—than about success in their attempt to organise a small data collection task together. After I had realised this, my attention eventually moved from my original research questions to questions and observations about the various ways in which collaborative projects that involve extra-academic experts can create societal impact (Koskinen under review). In BizTech we could have designed a project that would have informed the university management on a regular basis and possibly had an impact on the development of the ongoing structural reorganisation at the university. But that would have deeply affected the nature of the interview data we would have been able to gather. It might also have hampered our ability to reflect on the case now, and to compare it to similar organisational developments in other universities.

Collaboration between philosophers and scientists, or even the role of a consultant, or participation in science policy initiatives, can offer a philosopher of science highly effective ways to influence science or science policy (see, e.g., Keven et al. 2018; Vermier et al. 2018; Parker & Lusk 2019; Beck et al. 2021). If the aim is to have a relatively fast, straightforward impact, such approaches can be preferable to case study methods that require non-participation.

But the latter methods too can be well aligned with the normative aims of philosophy of science, and the wish to have an impact on science or on science policy. In the long run, they too can be quite effective. Of the two case studies I have just described, particularly in BizTech our results seem to be relevant to science policy. It could even be argued that our normative aims required that we keep a certain distance and do not offer comments or recommendations during the data collection. Our ability to produce results that are relevant in science policy more generally, not just at the university we were studying, depended partly on our methodological choices. As we did not attempt to influence the developments we studied, our results are more likely to be of interest when considering similar developments elsewhere.

Much of the literature on inter- and transdisciplinary research concentrates on examples where the researchers' own research interests have lead them to inter- and transdisciplinary collaborations (e.g., Hirsch Hadorn et al. 2008; Frodeman 2017). In BizTech, the inter- and transdisciplinary collaborations that emerged as a result of the structural reorganisation differed in several ways from the kind of inter- and transdisciplinarity that is typically described in the literature. Similar institutional and organisational changes than the one we studied are being implemented in many countries, and they are meant to encourage and incentivise scientists towards inter- and transdisciplinary collaborations. Our findings suggest that they may be producing something else than originally intended. (See Salmela, MacLeod & Munck af Rosenschöld 2021; see also MacLeod & Nagatsu 2018; Salmela & Mäki 2018; Lindvig & Hillersdal 2019.)

The methodological decisions philosophers of science make when planning case studies limit the ways in which they can influence the science they are studying. Therefore, such decisions must be made in light of the normative aims of the study. In philosophy of science, the decisions will differ from similar decisions in other fields, such as sociology of science or STS. Even when the methodological ponderings and pros and cons might be similar, the normative aims of philosophy of science can and should influence the decisions, and this may lead to different decisions than would be warranted in some other field. This is something I believe must be taken into account when planning collaborative projects with STS scholars or sociologists or science—our partially dissimilar aims can lead to dissimilar methodological choices. Moreover, this means that philosophers of science cannot simply adopt methods from the social sciences—we must also adapt them to our needs, and continue developing them.

7 Conclusions

In principle a philosopher of science who is conducting a case study on ongoing research is in an excellent position to have an impact on the science they are studying. As Plaisance, Michaud and McLevey (2021) emphasise, the most effective pathways to impact in philosophy of science are interpersonal interactions—it is through conversations and even collaborations with scientists and policymakers, rather than through publications in philosophy journals, that our work has an impact outside our own field. Spending months or years with a group of scientists gives ample opportunities for such interactions.

There are situations, however, where a philosopher of science conducting a case study will not want to seize such opportunities, or will hesitate when they emerge. I have described two examples where this was the case. In SocJournArt I might have been able to have a stronger impact on the research conducted and the results of the project than I eventually did, as the participants had little experience of collaborations between social scientists and artists, and were therefore willing to listen a philosopher of science. But I was cautious, because I had not planned to conduct an experiment on whether I would able to facilitate such collaborations. In BizTech our team might have been able to influence the development of the structural reorganisation we were studying. The reorganisation was seen as an organisational experiment, and both some members of the university management and people involved in research funding were interested in our findings. But we decided not to attempt anything of the sort, as both research ethical considerations and our epistemic and normative interests led to the conclusion that we should not try to influence the processes we were studying.

Using qualitative case study methods in philosophy of science can offer ways to have an impact on science, but the impact is not necessarily a direct one. As I have noted, there are methods and approaches—particularly common in fields like development studies—that allow engaging with and having an active impact on the processes and developments one is studying. But often it makes sense to conduct a case study where such impacts are avoided. For a philosopher of science this is a loss: the chosen methods limit our ability to take the role of the "voice of reflection" within the research group or organisation we are studying. Such limitations can, however, be worthwhile, if the knowledge and understanding gained in the case study is valuable enough from the point of view of the epistemic and normative aims of the philosophers of science involved.

References

Adler, C., Hirsch Hadorn, G., Breu, T., Wiesmann, U., & Pohl, C. (2018). Conceptualizing the transfer of knowledge across cases in transdisciplinary research. Sustainability Science, 13(1), 179–190.

Ankeny R. A., Chang H., Boumans M., & Boon M. (2011). Introduction: Philosophy of Science in Practice. European Journal for Philosophy of Science, 3(1), 303–307.

Beck, J. M., Elliott, K. C., Booher, C. R., Renn, K. A., & Montgomery, R. A. (2021). The application of reflexivity for conservation science. Biological Conservation, 262.

Boumans, M. & Leonelli, S. (2013). Introduction: On the Philosophy of Science in Practice. Journal for the General Philosophy of Science, 44, 259–261.

Brister, E. and Frodeman, R. (2020). Digging, Sowing, Building: Philosophy as Activity. In Brister, E. & Frodeman, R. (eds.) A Guide to Field Philosophy: Case Studies and Practical Strategies. Milton Park: Routledge, 1–14.

Brown, V. A., Deane, P. M., Harris John, A. and Russell, J. Y. (eds.) (2010). Tackling wicked problems through the transdisciplinary imagination. London: Earthscan.

Bursten, J. R. S. (2020). Lab Report: Lessons from a Multi-Year Collaboration between Nanoscience and Philosophy of Science. In Brister, E. & Frodeman, R. (eds.), A Guide to Field Philosophy: Case Studies and Practical Strategies. Routledge, 35–47.

Carew, A. L., and Wickson, F. (2010). The TD Wheel: A heuristic to shape, support and evaluate transdisciplinary research. Futures, 42(10), 1146–1155.

Chao, H.-K. & Reiss, J. (2017). Philosophy of Science in Practice. Nancy Cartwright and the Nature of Scientific Reasoning. Synthese Library, Vol. 379. Cham: Springer.

Choi, B. C. K., and Pak, A. W. P. (2006). Multidisciplinarity, interdisciplinarity and transdisciplinarity in health research, services, education and policy: 1. Definitions, objectives, and evidence of effectiveness. US National Library of Medicine National Institutes of Health, 29(6), 351–364.

Frodeman, R., Klein, J. T., & Dos Santos Pacheco, R. C., editors (2017). The Oxford Handbook of Interdisciplinarity. Second edition. Oxford: Oxford University Press.

Ginsberg, A. D., Calvert, J., Schyfter, P., Elfick, A., & Endy, D. (2014). Synthetic Aesthetics: Investigating Synthetic Biology's Designs on Nature. Cambridge MA: The MIT Press.

Hartley, J. & Benington, J. (2000). Co-research: A new methodology for new times. European Journal of Work and Organizational Psychology, 9:4, 463–476.

Hirsch Hadorn, G., Biber-Klemm, S., Grossenbacher-Mansuy, W., Hoffmann-Riem, H., Joye, D., Pohl, C., Wiesmann, U., & Zemp, E. (2008). The emergence of transdisciplinarity as a form of research. In Hirsch Hadorn, G., Hoffmann-Riem, H., Biber-Klemm, S., Grossenbacher-Mansuy, W., Joye, D., Pohl, C., Wiesmann, U, & Zemp, E. (eds.), Handbook of Transdisciplinary Research. Dordrecht: Springer, 19–42.

Hirsch Hadorn, G., Pohl, C., and Bammer, G. (2010). Solving problems through transdisciplinary research. In Frodeman, R., Klein, J. T. & Mitcham, K. (eds.), Oxford Handbook of Interdisciplinarity. Oxford: Oxford University Press, 431–452.

Huutoniemi, K., Thompson Klein, J., Bruun, H., & Hukkinen, J. (2009). Analyzing interdisciplinarity: Typology and indicators. Research Policy, 39(1), 79–88.

Jacobs, J. A. & Frickel, S. (2009). Interdisciplinarity: A critical assessment. Annual Review of Sociology, 35(1), 43–65.

Keven, N., Kurczek, J., Rosenbaum, R. S., & Craver, C. F. (2018). Narrative construction is intact in episodic amnesia. Neuropsychologia, 110, 104–112.

Koskinen, I. (2018a). Miksi tieteilijöiden kannattaa tehdä yhteistyötä taiteilijoiden kanssa. Ajatus, 75, 93–119.

Koskinen, I. (2018b). Että maailma muuttuisi paremmaksi. In Lähde, V. & Vehkoo, J. (eds.) Jakautuuko Suomi? Helsinki: Into Kustannus.

Koskinen, I. Under review. Societal impact in research collaborations across the boundaries of science.

Koskinen, I. & Mäki, U. (2016). Extra-academic transdisciplinarity and scientific pluralism: What might they learn from one another? The European Journal of Philosophy of Science, 6(3), 419–444.

Koskinen, I. & Rolin, K. (2022). Distinguishing between legitimate and illegitimate roles for values in transdisciplinary research. Studies in History and Philosophy of Science, 91, 191–198.

Kosolosky, L. (2012). Philosophy-of-Science in Practice vs. Philosophy of Science-in-Practice. Newsletter SPSP, Winter 2012, 9–10.

Leavy, P. (2011). Essentials of transdisciplinary research: using problem-centered methodologies. Walnut Creek CA: Left Coast Press.

Lindvig, K. & Hillersdal, L. (2019). Strategically unclear? Organising inter-disciplinarity in an excellence programme of interdisciplinary research in Denmark. Minerva, 57(1), 23–46.

MacLeod, M. & Michiru N. (2018). What does interdisciplinarity look like in practice: Mapping interdisciplinarity and its limits in the environmental sciences. Studies in History and Philosophy of Science, 67, 74–84.

Maassen, S. & Weingart, P. (2005). What's New in Scientific Advice to Politics? In Maassen, S. & Weingart, P. (eds.), Democratization of Expertise? Exploring Novel Forms of Scientific Advice in Political Decision-Making. Sociology of the Sciences Yearbook, 24, 1–19.

Maassen, S. & Lieven, O. (2006). Transdisciplinarity: a new mode of governing science? Science and Public Policy, 33(6), 399–410.

Mansilla, V. B., Lamont, M., & Sato, K. (2016). Shared cognitive-emotional-interactional platforms: Markers and conditions for successful interdisci-plinary collaborations. Science, Technology, & Human Values, 41(4), 571–612.

Parker, J. & Crona, B. (2012). On being all things to all people: Boundary organizations and the contemporary research university. Social Studies of Science, 42(2), 262–289.

Parker, W. S. & Lusk, G. (2019). Incorporating User Values into Climate Services. Bulletin of the American Meteorological Society, 100(9), 1643–1650.

Plaisance, K. S. & Elliott, K. C. (2021). A framework for analyzing broadly engaged philosophy of science. Philosophy of Science, 88(4), 594–615.

Plaisance, K. S., Michaud, J., & McLevey, J. (2021). Pathways of influence: un-derstanding the impact of philosophy of science in scientific domains. Synthese, 199, 4865–4896.

Pohl, C. (2011). What is progress in transdisciplinary research? Futures, 43, 618–626.

Pohl, C. & Hirsch Hadorn, G. (2007). Principles for Designing Transdisciplinary Research. Proposed by the Swiss Academies of Arts and Sciences. München: oekom—Gesellschaft für ökologische Kommunikation.

Pohl, C., Truffer, B. & Hirsch-Hadorn, G. (2017). Addressing wicked problems through transdisciplinary research. In Frodeman, Thompson Klein, & Dos Santos Pachecho (eds., 2017), 319–331.

Potochnik, A. (2018). How Philosophy of Science Relates to Scientific Practices. Blog post (Auxiliary Hypotheses, 30 August 2018).

Robinson, B., Gonnerman, C., & O'Rourke, M. (2019). Experimental Philosophy of Science and Philosophical Differences across the Sciences. Philosophy of Science, 86(3), 551–576.

Salmela, M., MacLeod, M. & Munck af Rosenschöld , J. (2021). Internally Incentivized Interdisciplinarity: Organizational Restructuring of Research and Emerging Tensions. Minerva, 59(3), 355–377.

Salmela, M. & Mäki, U. (2018). Disciplinary emotions in imperialistic interdisciplinarity. In Mäki, U., Walsh, A., & Fernández Pinto, M. (eds.), Scientific Imperialism: Exploring the Boundaries of Interdisciplinarity. Milton Park: Routledge, 31–50.

Schrögel, P. & Kolleck, A. (2019). The Many Faces of Participation in Science: Literature Review and Proposal for a Three-Dimensional Framework. Science & Technology Studies, 32(2), 77–99.

Turner, V. K., Benassaiah, K., Warren, S., & Iwaniec, D. (2015). Essential tensions in interdisciplinary scholarship: Navigating challenges in affect, epistemologies, and structure in environment-society research centers. Higher Education, 70, 649–665.

Wagenknecht, S., Nersessian, N. J., & Andersen, H., editors (2015). Empirical Philosophy of Science: Introducing Qualitative Methods Into the Philosophy of Science. Cham: Springer.

Vermeir, K., Leonelli, S., Tariq, A. S. B., Olatunbosun, S., Ocloo, A., Khan, A. I., & Bezuidenhout, L. (2018). Global Access to Research Software: The Forgotten Pillar of Open Science Implementation. Global Young Academy, German National Academy of Sciences Leopoldina.

Zierhofer, W. & Burger, P. (2007). Disentangling transdisciplinarity: an analysis of knowledge integration in problem-oriented research. Science Studies, 20(1), 51–74.

Historical epistemology as a meta-reflection between science and philosophy

Fabio Minazzi

Dipartimento di Scienze Teoriche e Applicate, Universitá degli Studi dell'Insubria, Via J. H. Dunant, 3, 21100 Varese, Italy

E-mail: fabio.minazzi@uninsubria.it

> *"The most incomprehensible thing about the world is that it is comprehensible."*
> Albert Einstein

1 Theoretical premise: critical rationalism and the teachings of Banfi

1.1 Kant and the discovery of the transcendental

The fine and acute scholar Mario Dal Pra once observed that speaking of the theory of reason developed by Banfi entails making reference to some of the "most solemn voices in the whole tradition of thought". In fact, in Banfi's masterpiece Principles of a Theory of Reason (hosted and published in 1926 in the collection directed and promoted by Banfi's mentor, Piero Martinetti), it is explicitly evident that Banfi's critical rationalism coincides "substantially with a unitary critical rethinking of Kantianism and Hegelianism".[1] Reference to Kant implies, of course, the reference to the "critical problem" especially addressed by Kant, with the critical warning, however, that "if the critical problem is the soul of Kantian philosophy, the discovery of the transcendental is the soul of that soul."

Transcendentality, therefore, as a discovery and critical-epistemological awareness that human knowledge never constitutes an absolute unveiling of reality as such, but rather consists, if anything, in the strenuous and never guaranteed conquest of an objective knowledge which is developed and established, to say it with Husserl, within a precise, always delimited and circumscribed, "ontological region", within which knowledge is constructed by intertwining the principles of pure rationality with the complex plane of experimental verification. From this hermeneutic perspective, the Kantian transcendental coincides exactly with the well-known "Copernican revolution" expressly thematised and claimed as its own achievement by epistemological criticism, since every "reality" to which a physical theory cognitively

---

[1] M. Dal Pra, *Kantismo ed hegelismo in Banfi* in Autori Vari, *Antonio Banfi (1886-1957)*, Reports of the conference *Antonio Banfi: le vie della ragione*, University of Milan, 28 February 1983, Edizioni Unicopli, Milan 1984, pp. 21–35; the quotations that appear in the text are taken, respectively, from p. 21, p. 23; p. 24, pp. 25–26. On the work of Dal Pra within the "Milan school", see *Mario Dal Pra nella "scuola di Milano"*, edited by F. Minazzi, Mimesis, Milan-Udine 2018.

Science's Voice of Reflection, edited by G. Heinzmann & B. Löwe.
Comptes Rendus de l'Académie de Philosophie des Sciences I (2022), pp. 75–109.

refers has never to do with a mythical unrelated and absolute reality (that is, free from any constraint), but is constructed - and constituted - within a precise and finite theoretical context, with respect to which knowledge is always structured in the light of certain experimental procedures of verification. In short, to put it differently, according to the approach of Kantian criticism, human knowledge is always and only constructed within precise theoretical and experimental constraints.

For this reason, the Kantian discovery of the transcendental implies a decidedly and programmatically anti-metaphysical position, by virtue of which human knowledge relinquishes its aspiration to be able to establish absolute and metaphysical knowledge, at the same moment when it instead gains an objective knowledge which proves to be such only and within the limits defined by a given theoretical apparatus and in dialectical connection with an equally defined and precise experimental apparatus. Just this phenomenal knowledge generates the possibility for human beings of achieving some objective knowledge through which they begin to know, in a finite and always partial way, the world in which they live. As Dal Pra writes,

> the discovery of the transcendental is in essence the discovery of reason itself; in fact it is not the world of knowledge grasped in its infinite contents, in the endless multiplicity of its data, but identified as the result of the working of the form of that structure of which the germ of reason itself seems to properly consist.

Which naturally leads Kantian critical rationalism along a very specific path, the one in which knowledge can only be configured as a task that is always open, critical and procedural, never definitive, programmatically anti-metaphysical precisely because it is able to rediscover an internal "critical metaphysics" constitutive of all objective knowledge.

1.2   How can we think about reason from a historical perspective?

For this theoretical reason, Banfi, explicitly referring to the Hegelian lesson, thinks that it is also necessary to have the ability to grasp and historically understand "also the universal principles" of the theory itself, since it is necessary to know how "to think of reason historically. If therefore reason is form according to Kant, it is also in Hegelian terms a structure constructed over time" (my italics). To understand this intrinsic dynamism of rationality, Banfi thus looks, with decidedly Hegelian eyes, at the fruitful Kantian transcendental dialectic, having the ability to understand how Kantian ideas do not represent in the least an object given and codified, but constitute "the line of a rational process", always open and integrable. If indeed Kantian ideas express, according to the classic and traditional Kantian formulation of the Critique of Pure Reason, "the aspiration to the totality of the conditions of a given conditioned", thus by configuring an evident metaphysical impossibility (which, in the illusory transcendental dialectic, ends up, in fact, passing off as "absolute" a knowledge that is in reality always circumscribed and finite), it is therefore necessary to associate, as can be deduced from the Hegelian lesson, the Kantian concept of limit with the idea itself. In this way, Dal Pra observed with great exactness,

> the concept of limits reinforces in a certain way the concept of idea, in the sense of opening it towards a reference to what goes beyond it; and if we take into account that already the concept of the idea does not represent an object, but 'the line of a rational process', the concept of the idea-limit reaffirms, so to speak, within the same line of the rational process, the reference to the further development of the process itself, its further tension. In short, the concept of the idea-limit strengthens and consolidates the process and removes any dogmatic limit from it.

In this way, the intrinsic critical processuality of knowledge is placed in the heart of Kantian criticism itself, making it possible to delineate a critical, problematic and open rationalism, which in this singular intertwining of Kantianism and Hegelianism, is actually capable of going beyond the lesson of the two great classic German philosophers, in order to delineate a new and more plastic, problematic, critical and hermeneutic horizon. Precisely this new and fruitful horizon constitutes, at the same time, the theoretical program[2] of philosophical, cultural and civil research inaugurated by Banfi's teachings in the context of the European culture of the first decades of the 20th century.

1.3   Banfi and the pure theoretical significance of knowledge

In this way the double critical fusion of Kantianism and Hegelianism successfully performed by Banfi in *Principi di una teoria della ragione* (*Principles of a Theory of Reason*) to outline his new critical rationalism, extends, as Dal Pra wrote,

---

[2]Regarding the critical use of the term "theoretical", often used by Banfi in a declaredly programmatic way, it should however be remembered that, not many years ago, there was a preliminary, dogmatic and programmatically uncritical resistance in the university of Milan often expressed with arrogance and remarkable verbal violence, by some exponents, then *à la mode*, of the so-called new epistemology (Lakatosian and/or Feyerabendian) of Popperian inspiration. According to them, in reality, there would not be any "theoretical" dimension because everything would be reduced only to the "theory". In which we can feel, already on a lexical level, the intrinsic theoretical poverty of these traditional "sunflowers of philosophy" (to say it with the philosopher Erminio Juvalta). Since I graduated in the early eighties of the last century with Giulio Giorello on the immanent procedural transcendence of knowledge, I have had to defend the permissibility of the usage of the term "theoretical" which was systematically dismissed and usually replaced (in a clearly erroneous way) with "theory"...

the horizon of reason beyond the limits marked by Kant, accentuating
its procedural disposition, beyond any closure, both psychological-
subjective and historical, and moreover in the sense of consolidating
its function and autonomy.

For this precise theoretical reason Banfi began in his Principles by stressing
that knowledge should be understood

> in its pure theoretical meaning, as mere knowledge, or, if we want
> to proceed to the determination and transcendental analysis of the
> idea of knowledge, as a law for which in every concrete cognition, the
> infinite task of theoreticality is immanent, as the synthesis of certain
> elements.[3]

For this same reason too, Banfi could then state that his theoretical research
is and remains authentically

> transcendental, and the actuality of knowledge, the ways of its con-
> crete determination in the plans of experience became for [him] a
> problem that presupposed the transcendental analysis of the idea of
> knowledge, but cannot be resolved by it, since for a solution it re-
> quired rather a previous recognition of the nature of theoreticality,
> its relationship with reality, and, specifically, with the spiritual reality
> to which facts and cognitive relationships belong.

The concept of the transcendental is therefore assumed here by Banfi in the
precise sense imposed and deployed by the famous "Copernican revolution"
inaugurated by Kant with the discovery of the transcendental as a "moment
of autonomous legitimacy which founds the unitary structure of experience
and is independent of its determined aspects."

But at the very moment when Banfi referred to the critical heart of Kantian transcendentalism, he nevertheless accentuated, as mentioned before,
"the transcendental analysis of the idea of knowledge itself", developing, on
the one hand, the typical direction of rationality and denouncing, on the
other hand and at the same time, the traditional dogmatism that absolutises
the different constituent moments of the transcendental structure. For this
reason, in Banfi's analysis, knowledge is

> considered and subjected to a transcendental analysis in its pure con-
> ception, with respect to the universal law according to which it dom-
> inates and give sense to the relationships and aspects in line with
> which it intersects with the reality of spiritual life.

---

[3] A. Banfi, *Principi di una teoria della ragione*, Editori Riuniti, Rome 1967, p. 8, while
the quotations that follow in the text are taken from the following pages respectively:
pp. 8–9; p. 11; p. 13; p. 19; p. 20; p. 21; p. 23; p. 40; p. 44.

The dual structure of the subject-object antinomic relationship itself, which structures the idea of knowledge, thus represents for Banfi not an original datum of consciousness, but a product of his own critical procedural investigation:

> the subject-object relationship is not given originally to consciousness; it develops rather and rises more and more clearly as the theoretical sphere and the cognitive activity gain autonomy in cultural self-awareness.

Also in this case Banfi is not interested in defining knowledge according to one of its different and multiple phenomenological positions, since his aim is, if anything, the opposite, to investigate and critically clarify knowledge by fully bringing out "its pure universal theoretical structure, its typical formal relationship", enabling us to understand how "the transcendental character of the subject-object gnoseological relationship, makes it absolutely correlative in nature", since "subjects and objects are not such according to their being determined, or to a content that characterises them, but only as a function of their correlation, of the unity itself which is precisely knowledge": "this pure relationship of subject-object correlation, this transcendental synthesis of the two terms constitutes the essential form or the idea of knowledge."

1.4   Banfi and the transcendental law of knowledge

In short, this correlation constitutes

> the transcendental law of knowledge, which dominates and directs the infinite process and development of the action of knowing". Therefore the subject-object gnoseological relationship does not constitute the expression of an empirical or metaphysical order, because, if anything, "in the transcendental synthesis of the two terms, for which these have their pure gnoseological value, it expresses the theoretical requirement that characterises knowledge and that constitutes the law of its development in its various aspects.

On the other hand, these two antinomical moments of knowledge, precisely because they always take root on an open plane of immanent procedural transcendentality, refer to a plurality of ideal correlations that

> can only be valid as transcendental moments. Transposed into reality, on this plane, their own ideal unity becomes the principle of their irreducible antithesis. This typical situation can be defined as the universal problematic of knowledge.

Consequently, Banfi's critical rationalism starts from the radical, inspirational, Kantian and Hegelian theoretical assumption, according to which "the problematic nature of every knowledge appears here as the law of its infinite development". This allows us to understand the intrinsic critical radicality of Banfi's perspective which broke with all absolute metaphysical rigidity, while re-proposing the intrinsic problematic nature of knowledge as such, underlining its intrinsic *Copernican* character, because within this critical-transcendental perspective it is no longer possible to "naively and metaphysically" explain the extent of knowledge on a plane of purported absolute domains. Banfi specifies that:

> the formula of truth as *adaequatio intellectus et rei* expresses precisely this need, but as long as the *intellectus* remains such and the *res* remains *res*, i.e., determined on the basis of extra-cognitive relationships, their relationship cannot be simplified in gnoseological *adaequatio*, that is, in the cognitive synthesis, but rather represents an unsolvable alterity. Moreover this *adaequatio*, which is inconceivable as long as the intellect and the thing are considered as concrete and absolutely determined aspects of reality and knowledge as a concrete relationship occurring between them, takes place in the process of knowing itself, as a transcendental synthesis, in which the two terms resolve, in the theoretical form, their being in themselves, in order to be valid as the two ideal poles, in whose relationship the cognitive relationship develops and the transcendental form of theoreticality extends to the whole content of experience.

In the cognitive relationship, the subject-object synthesis thus constitutes an ideal immanent law and an infinite term of a process that is always critically open. Kantian criticism, thanks to the Hegelian lesson, is therefore radically historicised and open to the processuality of historical knowledge, while, on the other hand, Hegelianism, thanks to the Kantian formalism concerning the transcendental structure of knowledge, is instead critically problematised on the level of mere formality which, in fact, structures every possible knowledge, every *logos*.

1.5 Banfi and the intrinsic problematicity of knowledge

In this theoretical perspective connected with the transcendental principle of knowledge, the two ideal poles of subject and object, of ego and of particular objects are not then taken

> as a fundamental dogmatic presupposition, but simply as they are, given relatively in experience. All knowledge presupposes precisely a being given of a mutual determination of the two terms: the ego and things are among themselves in a system of relationships that can be the system of physical reality or that of cultural reality, or rather it is both the one and the other at once, and in this intertwining they are mutually determined.

Therefore, at least in concrete and effective knowledge, variously codified within a specific and particular technical-cognitive heritage, the specific determination of the two correlated extremes of subject and object

> dissolves, and is in principle dissolved, and therefore concrete knowledge is the recognition and development of their full relativity, which in the theoretical sphere extends to its universal form.

The conclusion of this critical-rationalist approach can only be identified in the underlining of the *intrinsic problematicity of knowledge as such*:

> The problematic nature of knowledge thus expresses, in each particular act of knowing, the immanent transcendentality of the theoretical synthesis, which turns knowledge into an infinite process and does not allow it to stabilize and exhaust itself in a particular relationship between aspects determined by experience. And, precisely because this problematicity does not allow the determined positions of reality, and their partial and determined relationships, to be considered as absolute, it is the formal condition for highlighting the complexity of the relationship structure of reality and this results in a system of relationships theoretically detectable. In other words, this problematic nature of knowledge is the only guarantee of the universal development of the theoretical sphere, because every limitation to the theoretical sphere is stated as problematic, as a function of particular data of experience.

2   The neo-positivist epistemology and its image of rationality

If one considers the overall epistemological debate of the twentieth century, one cannot deny that the tradition of logical empiricism, which arose firstly from the lesson of the *Wiener Kreis* developed from the teachings of Moritz Schlick and of his best known and most valid collaborators (from Rudolf Carnap to Friedrich Waismann, from Otto Neurath to Hans Hahn and Philipp Frank, not to mention, then, the original position of Hans Reichenbach, etc.), ended up largely by characterising the philosophical reflection on science in the past century.[4] As is known, the original Viennese

---

[4]In this regard, we can naturally think of the classic Viennese "manifesto" of neo-positivism, *The Scientific Conception of the World: The Vienna Circle* (Ernst Mach Society, 1929) authored by Hahn, Neurath and Carnap et al. (Italian edition edited by Alberto Pasquinelli, translated into Italian by Sandra Tugnoli Pattaro, Laterza, Rome-Bari 1979), which can be read together with the interesting and emblematic text by Moritz Schlick, *The Vienna School and Traditional Philosophy*, in Moritz Schlick, *Philosophical Papers*, Vienna Circle Collection 11/II, edited by H.L. Mulder and B. F. B. van der Velde-Schlick, Reidel, Dordrecht, Boston and London (1979) pp.491-498. Italian version: *La scuola di Vienna e la filosofia tradizionale*, curated by Federico Filippo Fagotto, La Tigre di carta-La Taiga, Milan 2019, which helps to better position the neo-positivist research program in relation to the previous western philosophical tradition.

approach owed much, in its turn, to the *Tractatus Logico-Philosophicus* by
Ludwig Wittgenstein, which in the Viennese context, however, was largely
misinterpreted. The basic intent of Wittgenstein's powerful masterpiece
was in fact rooted in a very precise ethical (and metaphysical) conception
that its author expressed well in the seventh proposition, which symbol-
ically concludes the *Tractatus*: "Whereof one cannot speak, thereof one
must be silent." Precisely in relation to this closing sentence, which was
both peremptory and emblematic, the neo-positivists ended up by misinter-
preting its metaphysical and philosophical meaning. In fact, the Viennese
thinkers believed that with this affirmation Wittgenstein wanted to abolish
and also cancel the possibility itself of referring to the ethical, moral and
metaphysical dimension which had to be necessarily confined to the context of the "silence" with respect to which, in fact, we cannot say anything.
And precisely for this reason from their neo-positivist perspective it had to
represent a secondary dimension to be neglected programmatically. Thus
of the *Tractatus* the neo-positivists retained above all that authentic "war
machine" with which Wittgenstein had divided sentences into two classes:
the significant and the insignificant (coinciding with pseudo-sentences). The
former, as is well known, were divided, in turn, into two other subsets: on
the one hand, the one formed by analytic propositions (typical of logic,
mathematics, algebra and, more generally, of all disciplines based on de-
ductive inferences) which were reduced precisely to tautologies which were
true by virtue of their logical form; on the other hand empirical or synthetic
propositions that to be true, since they refer to the world, must undergo
a rigorous verification process capable of confirming them and, precisely,
of "verifying them". The famous verificationism of the Vienna Circle, a
scandal and torment for the classical metaphysical tradition, precisely because it was presented by the Viennese neo-positivists as a sort of actual
"philosophical club" with which one could quickly silence any other metaphysical tradition (as well as any potential opponent to neo-positivism) is
rooted in the "epistemological war machine" built by Wittgenstein in his
Tractatus, which for this very reason was then revered by Viennese thinkers
as their true Bible. However, as Wittgenstein himself also came to realize at a certain point, things seemed much more complicated and intricate
than the "happy" epistemological war machine of the early Viennese neo-
positivism suggested. From this particular point of view the history—truly
emblematic—of logical empiricism, considered in all its phases and in all
its very fruitful movements, would finally mature in the "American phase"
of this movement of thought, which would find its emblematic expression
in the *International Encyclopedia of Unified Science*, (Chicago, 1938). But
the history of this movement of thought can be correctly made to coincide
precisely with the three different formulations of the verification princi-

ple formulated by neo-positivists.[5] Logical empiricism thus passed from a "narrow" formulation of the verification principle which distinguished the *Wiener Kreis* of 1928, to his first "liberalisation" which coincides substantially with Carnap's physicalism (1936-37), and then it reached the phase of "broad" empiricism, which characterised the American developments of this movement during the post-war period. Through this fruitful development of continuous critical analysis

> logical empiricism snapped many hoops of the metaphysical barrel in which it had been sealed by the positivism originating with Mach or Russell or Wittgenstein. The relation between theoretical discourse and experience became more dynamic and therefore more fluid: the fruitful tension between syntax and factuality, which constitutes the truly progressive element of science, received its first recognition in theory; it paved the way for the elaboration of the logical techniques for the languages of the empirical sciences.

In other words, with the strict rethinking of the verification principle, the neo-positivists begin to realise, epistemologically speaking, that between heaven and earth there are more things than they initially imagined. If in fact in the initial phase of the Wiener Kreis they had been lulled in the (metaphysical!) dream that all scientific statements could be reduced, without unaccounted residues, to the verified "facts", in the second half of the 1930s, neo-positivists began to realise that the relationship between statements and verification is more complex than they had previously imagined. In the American phase, this critical awareness is articulated even further without, of course, ever abandoning the privileged empiricist horizon of reference. In this regard, Preti rightly observed that

> the new logical empiricism responds by further 'liberalising' the preceding empiricism making it even more markedly empirical. It takes to an extreme the dual conception whose seed had been sown by Reichebach and which Neurath had also glimpsed in his dispute with Schlick. Every scientific discourse consists, or may consist when it achieves a certain ideal of rigour, of an uninterpreted system of deductive symbols and an interpretation that establishes a correspondence, or rather an association, of at least some statements of the theory (which could be taken as the set of the consequences of the theory), and therefore usually of at least some concepts of the formal system. These concepts are normally not primitive but complex

---

[5]From this point of view, the critical examination carried out by Giulio Preti in his acute essay is still emblematic. *Le tre fasi dell'empirismo logico (The Three Phases of Logical Empiricism)* appeared originally in Mario Dal Pra's journal "Rivista Critica di Storia della Filosofia" (year IX, January–February 1954, fasc. I, pp. 38–51) and subsequently re-issued in G. Preti, *Philosophical Essays*, edited by Fabio Minazzi, translated by R. Sadleir, Peter Lang, Brussels, 2011; quotations appearing in the text are taken respectively from pp. 176–177, from p. 177 and from p. 178.

conceptual formations within the system, with empirical statements, in their turn consist both of predicates of observation and empirical constructions (protocols). What in this way becomes 'testable' is the whole system: its fertility in explanations, applications, forecasts— in a word, its operativity, by which the system itself stands or falls. Needless to say, no system is absolute or definitive. The empiricist is well aware of this, and his concern is to ensure the indefinite progressiveness of knowledge rather than to attribute some supposedly absolute value to it.

In short: in the first phase of the Vienna Circle, for neo-positivists a statement makes sense if and only if, in principle, it is verifiable. In the second stricter phase, an utterance turns out to be meaningful if and only if, in principle, it is interpretable, that is translatable, through some operations, into an observational statement. Thirdly and finally in the American "liberalised" phase of neo-positivism, it is stated that a theory has its own factual sense if and only if, again in principle, a particular set of statements (that is, the set of its consequences) is associated with a set of observational statements. Naturally between these three different formulations of the verification principle there are also precise logical connections, since the third formulation implies the previous two as its particular cases, while the second version also includes the first formulation as a much more delimited and circumscribed case. From this point of view, observed again Preti,

> we have come to distinguish *more or less* three degrees of *empirical certainty* which in some sense parallel the 'degrees of rigour' that some contemporary mathematical currents distinguish in mathematical propositions.

But, Preti adds conclusively,

> note that in spite of the empirical requirement, our discourses should ultimately relate to empirical observations and operations, and that they find only in them any value as factual knowledge—this requirement, I say, remains unchanged through all these phases. By giving way to these enlargements of the field of admissible scientific discourses, empiricism has not denied itself—it has only become gradually more ... empiricist.

3   Hempel and the epistemological dissolution of logical empiricism

However, it could be objected, this sort of fruitful and progressive "critical maturation" of empiricism was achieved also by challenging the "metaphysical nucleus" of the great tradition of empiricism, that is, its utopian desire of being able to reduce, *without residues*, the theoretical statements on a factual plane. Certainly the reference to the dimension of actuality is always very present - and on this point Preti is completely correct to observe how logical empiricism became increasingly ... empiricist; however, the empiricism we are now considering was profoundly different from the factual horizon to which the *Wiener Kreis* initially referred. But in what was it different? Exactly in the awareness that the mythical verification-ist principle, which initially was employed to attack traditional philosophy and, in particular, metaphysics is, actually, mythical. Against this mythology of empiricist derivation, a very different and much more sophisticated epistemological and philosophical awareness was developed. This was well expressed by Carl Gustav Hempel, who in *Fundamentals of Concept Formation in Empirical Science* (1952) and *The Theoretician's Dilemma* (1958) has managed to understand

> concept formation in science cannot be separated from theoretical considerations; indeed it is precisely the discovery of concept systems with theoretical power which advances scientific understanding; and such discovery requires scientific inventiveness and cannot be replaced by the certainly indispensable but also doubtless insufficient—operationist or empiricist requirement of empirical import alone.[6]

In fact, thanks to this mature reflection by Hempel, the tradition of logical neo-positivism came to unveil the fundamentally twofold nature of the scientific undertaking, fully realising—on a purely epistemological level—

> that an interpreted scientific theory cannot be considered equivalent to a system of propositions, whose extralogical constituent predicates are all either observational terms or obtainable from such predicates through reductional propositions: *a fortiori*, no scientific theory is equivalent to a finite or infinite class of statements describing potential experiences.

In short: science must naturally take into due and fundamental consideration the empirical dimension and the horizon of experimental verification as well as that of its possible experimental falsification. However this level, although indispensable, does not completely explain the intrinsic complexity of the scientific enterprise which, to unfold and develop, it also requires *ideas, thoughts* as well as the ability to know how to build abstract theories through which we are able to try to improve the knowledge of the world in which we live. Which, however, after more than three centuries of almost undisputed epistemological domination of modern empiricism (from Hume's formulation to the neo-positivists' development) leads us, paradoxically, to rediscover the Galilean roots of scientific knowledge that Galilei had

---

[6] Carl G. Hempel (English edition), *Fundamentals of Concept Formation in Empirical Science*, University of Chicago Press, Chicago 1965, p. 47, while the quotation which follows in the text is taken from p. 37.

well identified and clarified in his methodological masterpiece, *The Assayer* (1623) in which he rightly insisted on highlighting how scientific knowledge arose from the critical intertwining of "sense experiences" and "necessary demonstrations".

Exactly within this complex and articulated dual perspective, within which the technical-experimental dimension always plays its own precise and indispensable role, scientific knowledge is thus built, which, if it cannot disregard the experimental verification or falsification procedures, on the other hand also needs the ability to build theories *ex suppositione* precisely because the mathematical scientist (*filosofo geometra*),

> must always be capable of 'deducing' the material hindrances, but to do so he must also be able to *think* the world by building scientific theories which enable him to discern significant aspects of a reality which, in itself, has also its own specific 'deafness' that critical intelligence must know how to penetrate in a fruitful way.[7]

Hence from this critical perspective the continuous development of the principle of empirical verification produced by logical neo-positivism can also be configured as a process by which this tradition of thought, as it managed to elaborate an increasingly sophisticated and critical epistemological reflection, compromised, however, the very foundations of its epistemological research programme. In this way, paradoxical as it may seem, the progressive critical maturation of logical neo-positivism ended up by coinciding with its own self-dissolution. In other words, it is precisely the underlying theoretical honesty of this movement of thought that ultimately determined its overall disappearance from the horizon of contemporary philosophical reflection. For what reason? Precisely because, as has been mentioned, this movement, by elaborating three different increasingly critical and sophisticated formulations of its verification principle, finally came to understand—through Hempel's reflections—that the verification principle itself, which was the fundamental tool for grasping the very essence of the scientific enterprise proved to be a blunt instrument. And albeit not useless, however, it required a profound change in its epistemology. Within this dramatic, purely theoretical (and constitutive) dilemma, neo-positivism thus ended up by dissolving itself at the very moment when it comprehended and criticised the limits of its own innovative and original research programme. This is also what constitutes, of course, the nobility and the undoubted theoretical greatness of this movement of thought, which constantly analysed, in depth and critically, its own point of view, and finally developed

---

[7]On the complex and articulated epistemological conception of scientific knowledge developed by Galileo Galilei, I may be permitted to refer to my volume *Galileo "filosofo geometra"*, Rusconi, Milan, 1994, in which I have analytically discussed many Galilean pages in which Galilei shows that he devised a sophisticated critical-epistemological vision of our knowledge of the world.

also the theoretical power to dissolve it in order to recognise the specific complexity and autonomy of the cognitive problem addressed. If we now look at the whole extraordinary critical parable—both from a historical and a theoretical point of view—of neo-positivism, we cannot, however, avoid asking a decisive question: what was the idea of rationality adopted by the philosophers of the Vienna circle? A question that naturally leads us to face the same problem also with regard to the reflection of Wittgenstein and Russell. Now, considering only the *Tractatus*, one has to investigate what conception of rationality Wittgenstein defended and proposed within the theoretical construction of his work, whose qualifying theses are almost "nailed" (almost "by oracular force") to the admirable overall texture of his masterpiece. Well, if we approach Wittgenstein's work from this particular point of view, it is easy to understand how the author of the *Tractatus* leaned towards a substantially algorithmic image of rationality. An algorithmic image of rationality which systematically reduced it to the formal and specific dimension of the logical form of tautology. In this logical-mathematical view of the rationality clearly derived from Russell, what is absolutely lacking is precisely the intrinsic plasticity of human reason. There is really no trace of this plasticity in the *Tractatus*, which, consequently, reflects a deeply weakened and impoverished idea of human reason, so much so that for Wittgenstein in science there is nothing mysterious, complex and extraordinary, since in his opinion in the field of science if a problem can be posed then its solution must necessarily be found. As Wittgenstein himself wrote in proposition 6.5 of the *Tractatus*

> for an answer which cannot be expressed the question too cannot be expressed. The *riddle* does not exist. If a question can be put at all, then it can also be answered.

For this precise reason, Wittgenstein also declares (in proposition 6.52 of the *Tractatus*):

> We feel that even if *all possible* scientific questions be answered, the problems of life have still not been touched at all. Of course there is then no question left, and just this is the answer.

An answer which therefore leads us beyond language, to the area of "silence" dominated by the awareness, as we have seen, that "Whereof one cannot speak, thereof one must be silent." (Proposition 7). And indeed, for Wittgenstein "Not *how* the world is, is the mystical, but *that* it is." (6.44).[8] In this way, scientific rationality is separated from a purely instrumental and almost "trivial" function, because human rationality is only

---

[8]L. Wittgenstein, *Tractatus Logico-Philosophicus*, Kegan, Trench, Trubner & C., London, 1922, pp. 89–90, the italic in the text is always Wittgenstein's.

concerned with understanding *how* the world is structured, while mystical reflection points to a much higher and unfathomable goal, the one that most directly concerns the *existence* itself of the world. The pictorial theory of language developed by Wittgenstein in the *Tractatus* constitutes a confirmation of a unidimensional image of rationality that reduces it to its merely formal dimension, depriving it of any plasticity and even of any creative originality. In fact, in Wittgenstein's work, as well as in that of his faithful Viennese "followers", this merely formal and "empty" image of human rationality emerges. As in the game of chess, the meaning of each piece is reduced to its legitimate moves, in a similar way for all these authors human reason is only a powerful algorithmic tool of inference and nothing more. Thus, while a scientist like Galileo was well aware how mathematics was able to "give wings" to human thought, critically opening up knowledge of spaces and dimensions never before imagined, for these authors it is precisely this sort of springing creativity of thought (also of mathematical thought) that is denied, precisely because they can see only the operational, algorithmic, functional and "mechanical" aspect of human reason. Thus, while Galileo, in his famous initial lines of his *Discourses and Mathematical Demonstrations Relating to Two New Sciences* openly polemicised against a traditional alienating vision of mechanics that systematically reduced it to a dimension devoid of any "creative spirituality" and even devoid of any "fruitful originality", these authors in line with Wittgenstein, Russell and the neo-positivists, ended up by subscribing to a weakened, formalistic and empty interpretation of human reason. If a great mathematical logician like Leibniz still perceived the power of the form, for these twentieth century authors this dimension was instead hopelessly lost just because they saw mathematics only from an algorithmic and technical perspective which deprives it, continuously and paradoxically, of any conceptual dimension. In this sense, these authors were then victims, paradoxically, of a mathematical formalism which progressively removed from mathematics any authentically *conceptual* dimension.

4 Verificationism and falsificationism: two sides of the same coin?

While neo-positivism carried out, with great "organisational" spirit, its fruitful research project, exercising its undoubted and significant hegemony, both in Europe and internationally (partly because of the Nazi occupation of Europe which forced many scholars to emigrate to the United States), however, there were some other authors in the field of epistemology, connected with a very different tradition of thought, who were capable of outlining a different and alternative idea of scientific knowledge and technical-scientific research. However, these different voices remained very isolated or (and at the same time) did not have the "organisational" capacity to create a sort of "common front" to defend and develop a different critical examination of the scientific enterprise. In this context I do not intend to refer particularly to Karl Popper's falsificationist epistemology which, too, was created in Vienna, with the publication, promoted directly by the *Wiener Kreis*, of his masterpiece *Logik der Forschung* (1934) in an editorial series directed by Moritz Schlick. I will not focus on this epistemological current for several reasons. *Firstly*, because the international resonance of the approach of falsificationism materialised itself only after the end of WW2, in the middle of the Cold War, when the "political" Popper was clearly used by Western forces in order to have an important liberal thinker who could convincingly oppose the Marxist tradition defended by eastern countries such as the USSR, the pivot of the socialist bloc. It is not surprising that it was the "political" success of Popper as a "philosopher of politics" and a staunch defender of the Western liberal tradition, the author, in particular, of *The Open Society and Its Enemies and The Poverty of Historicism*, which undoubtedly helped or facilitated the republication of his epistemological masterpiece which, not surprisingly, was subsequently reissued in a new English edition which appeared with the slightly modified title of *The Logic of Scientific Discovery* published in 1959, when Popper was teaching at the *London School of Economics*. *Secondly*, on a more strictly epistemological level, Popperian falsificationism—beyond what Popper himself claimed (he loved to present himself as the "killer" of neo-positivism), in reality owes much to the Viennese epistemological approach, to which it is linked by various features. The principal of these is his radical insensitivity to the history of science, which in his reflection always had an eminently "auxiliary" role in relation to epistemology. So if the Viennese neo-positivists wanted to find a definitive definition of science that would be able to explain, once and for all, the very "essence" of science as such (therefore considering it as completely detached and separated from the history of science), Popper also shared the same myth, since he was totally convinced that his falsificationism offered, finally, the real and authentic solution to the same problem, exquisitely epistemological, posed forcefully by neo-positivists. *Thirdly*, it cannot be ignored, that both in the Viennese verificationist reflection of neo-positivists and in the Popperian falsificationist one, no attention was ever paid to the problem, role and epistemological function of technology and technologies within the scientific enterprise. In this way, if one critically distances oneself from the idelological "trap", (inspired by pure "epistemological propaganda"), of the open opposition between neo-positivist verificationism and Popperian falsificationism, in reality, their deep (and

tacit) correspondence can be recognised, rooted, as it is, precisely in the
peculiar philosophical culture of the "Greater Vienna" in which both these
epistemological theories actually matured.[9]

5 Bachelard and a new conception of the activity of reason

Therefore, if we analyse the European epistemological debate, leaving in
the background both the neo-positivist and the falsificationist movements
(which in any case was "fruitful" only and solely after the end of WW2,
for the reasons already mentioned), some traditions of thought can be outlined that coalesced around authors who in those same years started some
interesting and original investigations of scientific knowledge. In this perspective we could mention the work of the Italian Federigo Enriques, or
that of Gaston Bachelard in France, or, again, that of Ferdinand Gonseth in Switzerland and also the particularly remarkable output of Ludwik
Fleck in Poland. It is, of course, not possible here to present the whole of
this articulated framework from which, however, the presence of different
voices and different traditions of thought emerged, which had the merit of
underlining some original or completely neglected aspects of the scientific
enterprise, developing perspectives for research that are still fecund and rich
in different results. Since I find it impossible to outline this general European framework (which is still under-researched), therefore I will focus, in

---

[9]In this regard, I would like to refer to my essay *Popper neopositivista deteriore?*
published in the volume written by various authors, *Riflessioni critiche su Popper*, edited
by Daniele Chiffi and Fabio Minazzi, Franco Angeli, Milan 2005, pp. 43–81, without
however neglecting one of the very first critical reviews of *Logik der Forschung*, i.e that by
Ludovico Geymonat published in his well-known *Logica e filosofia della scienza*, "Rivista
di filosofia", 3, 1936, pp. 250–265, in which the young Geymonat, also employing a precise
critical suggestion communicated to him by letter by Moritz Schlick himself, highlighted
a *constitutive fallacy* in Popperian falsificationism, since, even admitting the existence of
an asymmetry between verifiability and falsifiability, one can always argue, to put it in
the words of Popper himself (*and it was* 1959!), that it is still impossible,

> for various reasons, that any theoretical system should ever be conclusively fal-
> sified. For it is always possible to find some way of evading falsification, for
> example by introducing ad hoc an auxiliary hypothesis, or by changing *ad hoc* a
> definition. It is even possible without logical inconsistency to adopt the position
> of simply refusing to acknowledge any falsifying experience whatsoever. Admit-
> tedly, scientists do not usually proceed in this way, but logically such procedure
> is possible; and this fact, it might be claimed, makes the logical value of my
> proposed criterion of demarcation dubious, to say the least. (K. Popper, *The
> Logic of Scientific Discovery*, Routledge, London, 2002, pp. 19–20).

For the correspondence of Geymonat with Schlick in which we can read this interesting
letter by the founder of the *Wiener Kreis*, see my volume *Ludovico Geymonat epistemol-
ogo. Con documenti inediti e rari (un inedito del 1936, il carteggio con Moritz Schlick,
lettere con Antonio Banfi e Mario Dal Pra)*, Mimesis, Milan-Udine 2010, passim. Last
but not least I would like to mention the beautiful little volume by Geymonat should not
be forgotten, *Riflessioni critiche su Kuhn e Popper*, Dedalo Edizioni, Bari 1983.

particular, on the epistemological work (produced during his daytime activities!) of a fascinating thinker: Gaston Bachelard. Bachelard made his debut in the world of studies with an extraordinary book, the *Essai sur la connaissance approchée*, (published by Vrin, Paris, 1928), which even in the title stands out for its epistemological originality. The publication of this work constituted a sort of "meteorite" that appeared, quite suddenly, in the context of the philosophical and epistemological debate of the time. The title reveals the apparent "anomaly" of this new and unusual examination of the scientific enterprise. If, in the common perception, scientific knowledge is always seem as endowed with an almost absolute and undisputed rigour, on the contrary Bachelard instead wished to underline precisely the "approximate", precarious, always critically integrable, nature of scientific knowledge, making of *approximation* the very foundation of scientific knowledge. Which, of course, implied a radical reversal of some consolidated (and dogmatic) epistemological "commonplaces" (belonging, therefore, not just to common sense).

In the final pages of this book Bachelard stated, with a naturalness derived from his laboratory research, that in his opinion "approximation is the only fecund movement of thought",[10] precisely because he understood how the increase in human knowledge follows a growth curvature not unlike the one achieved by a vegetable during its development. Indeed, Bachelard writes:

> Let us consider life in its most distant and simplest form, that of the vegetable. We will notice that this kind of life achieves its adaptation only by somehow increasing its energy in an inventive and necessarily unexpected effort. Dr. Devaux points out the eminently active nature of mutations. Their origin 'would be due to a simple reaction of a plant when it is placed in the imperative condition of acclimatisation. This reaction is also active, which means that a plant, just like an animal, can occasionally free itself from the tyranny of the environment: and new acquired characteristics will be stable and hereditary precisely because they are not results imposed by the environment; this is equivalent to saying that all truly acquired characters are conquered characters'. Life, and perhaps all reality, is a progressive conquest of freedom. Its evolution adopts the very principle of rectification; in the assimilation, it accumulates the infinitely small advantages developed by the already realised organization: it deforms without breaking the shape; it normalises the accidental.

This attention by an epistemologist to the plant world certainly does not

---

[10]G. Bachelard, *Saggio sulla conoscenza approssimata*, translated and edited by Enrico Castelli Gattinara, Mimesis, Milano-Udine 2016, p. 269, while the other quotations that follow in the text are taken from pages: p. 279, p. 287; p. 48; p. 50; p. 51; p. 54; pp. 54–55.

constitute a very common stance, also because Bachelard looks at a discipline such as botany, to which the most committed neo-empiricist epistemologists would certainly prefer the hard sciences, i.e., physics and mathematics in the first place. Bachelard's unusual and important perspective reflects, moreover, his self-education, when he started teaching mathematics and physics in high schools, after having lived for some years as a post-office employee and worked in laboratories for years, accumulating a great and rich experimental experience that convinced him that the cognitive process, rather than being the result of a brilliant insight, *à la* Kuhn (which is said to arise suddenly, in the middle of the night),[11] is, if anything, the result of a minute, partial and continuous work, within which the knowledge of the world is built up slowly through an almost uninterrupted succession of continuous rectifications which assimilates the various elements within an uninterrupted adjustment. Exactly as happens in the plant world, where the growth of a plant presents a morphological development that arises precisely from this slow, tenacious and constant, continuous "adaptation" to an environment that in this way is originally and creatively "built" and variously "shaped" by the plant. Plants, in fact, in their very long evolutionary history, not only constantly adapted themselves to their environment, but built and shaped it creatively. Over the four billion years of their existence they have shown that they were capable of surviving different mass extinctions, from which they have always emerged with renewed vitality. Furthermore, Bachelard writes:

> How can we not be struck by the rectifying trend of a thought? Nothing is clearer and more fascinating than this conjunction between the old and the new. Rectification is a reality, or rather it is the real epistemological reality, because it is thought in its act, in its profound dynamism. Thought cannot be explained through the inventory of its acquisitions, because a force runs through it that must be accounted for. On the other hand, a force is well explained by indicating its meaning, its purpose. The goal to which the experimental determinations aim can be stated already when they apply to the scheme of an approximation. Approximation means unfinished objectification, but it is a prudent, fruitful, truly rational objectification, because it is aware at the same time of its own insufficiency as well as of its progress.

Rectification therefore proceeds just like a plant which, instant after instant, assimilates and transforms inorganic matter creating a new reality based on

---

[11]Thomas S. Kuhn, *The Structure of Scientific Revolutions*, The University of Chicago Press, Chicago & London, 2012, p. 90: "The new paradigm, or a sufficient hint to permit later articulation, emerges all at once, sometimes in the middle of the night, in the mind of a man deeply immersed in crisis."

life, while always remaining faithful to itself and to the tenacity of its own growth which is exercised *within its limits*.

Conceiving scientific knowledge as an "indefinite rectification" Bachelard not only showed that he was well connected with the actual development of experimental research, as it is carried out in every laboratory, but introduced into the very heart of knowledge that *intrinsic historical dynamism* of knowledge that the other epistemological currents (one need only mention neo-positivism and also falsificationism) never gave due consideration or that they certainly marginalised, pursuing the mythical objective of being able to define, once and for all, precisely *unhistorically*, the supposed and mythical immutable essence of science, the "quiddity" of science as such. On the contrary, for Bachelard "the differential equation of the epistemological movement" is provided precisely by the "continuous rectification of thought in the face of reality", which constitutes, as he himself programmatically declared in the first chapter of his work, "the only true subject of this book". In this perspective, "functional assimilation, which is the most indisputable principle of evolution, in short, continues its work in utilitarian knowledge. In its deepest sense, rectification perfectly matches the progress of this assimilation. It must face the future by slowly flexing the past. At the root of the concept there is therefore an adaptable life, capable of preserving and capable of conquering. Knowledge, grasped in its lower dynamism, already implies an approximation in the process of improvement." If we then proceed to higher levels, it is easy to realize how "functional assimilation is thus continued by intentional assimilation, that is to say, by an active choice."

This enabled Bachelard to highlight the decisive role that the *conceptual dimension* always carries out within scientific knowledge: "the concept, which is the element of a construction, has its full meaning only within the construction itself; and it is through a proposition that it is possible to naturally express the minimum knowledge of which it can be the object." Bachelard's insistence on concepts is also important and decisive, because it places his philosophy of science on a quite different and alternative epistemological position than that of the tradition of modern and contemporary empiricism. As we have seen, this great tradition of thought in fact pursued an unattainable utopia, namely that of being able to reduce, *without any residue*, knowledge to the factual dimension. On the contrary, Bachelard realised instead that scientific knowledge is always rooted in a specific and peculiar *conceptual dimension*, through which the real—continuously adjusted by continuous approximations—is precisely "conceived", i.e., transformed into conceptual reality. Einstein himself defined scientific knowledge just as "the mental grasp of this extra-personal world"[12] and with this ex-

---

[12] "Out yonder there was this huge world, which exists independently of us human beings and which stands before us like a great, eternal riddle, at least partially accessible

traordinary expression he managed to express, in an admirable way, the decisive role and heuristic function performed by the conceptual dimension within scientific research.

This decisive and fundamental *conceptual dimension* was instead systematically removed and never taken into due consideration by the epistemology of empiricist and verificationist theorists (as well as by falsificationism). Showing an evident Husserlian phenomenological influence, Bachelard distinguished "predicates from the act that unites them" and observed that

> the fact of determining as a subject a coherent synthesis of predicates is no longer attributable, according to an inverse analysis, to the knowledge of the attributes separated from each other. The synthetic judgment that defines a concept must avoid tautology, otherwise there would not really be any synthesis.

In disagreement with Wittgenstein and also with the *Wiener Kreis*, the conceptual dimension of scientific knowledge thus became the privileged terrain in which it is possible to achieve that continuous rectification of thought that allows us to build an approximate knowledge of the world and reality. While for the traditional verificationist epistemology (and the same observation also applies to the falsificationist epistemology) the famous Newtonian expression that force equals mass times acceleration was interpreted as the expression of a formula that summarizes, in universal and necessary terms, an almost infinite number of experiences experienced (and experimentable), on the contrary for Bachelard, $f = m \cdot a$ translated and constituted a specific *conceptual approach* from which a determined and circumscribed objective "approximation" of the world can be developed, which, thanks to its heuristic mediation, we want to get a knowledge of. In this new and original Bachelardian epistemological perspective "its definition, when actually conceived, is the translation of a real epistemological movement". In any case, Bachelard further explained, "if we consider knowledge in its full endeavour, we must always consider concepts as developed on a synthetic judgment in action".

---

to our inspection and thinking. The contemplation of this world beckoned as a liberation, and I soon noticed that many a man whom I had learned to esteem and to admire had found inner freedom and security in its pursuit. The mental grasp of this extra-personal world within the frame of our capabilities presented itself to my mind, half consciously, half unconsciously, as a supreme goal. Similarly motivated men of the present and of the past, as well as the insights they had achieved, were the friends who could not be lost. The road to this paradise was not as comfortable and alluring as the road to the religious paradise; but it has shown itself reliable, and I have never regretted having chosen it." (Albert Einstein, *Autobiographical notes*, Open Court Publishing Company, La Salle, Illinois, 1996, p. 5.)

6   Bachelard's dialectical "suprarationalism"

But what is then the characteristic of a concept according to Bachelard? In his view, "a concept is in fact an arrest [*arrêt*] in analysis, an actual decree by which the features outlined for a given object are considered sufficient to recognise it". An epistemological analysis must naturally always consider this characteristic of concepts that "arrests" our own possibility of thinking about reality, and must do so by always paying attention to the interlocking nature of the scientific knowledge of the world, contemplating carefully the two different poles within which this knowledge is always built: "on the one hand things with their more or less visible differences, on the other hand the spirit with its discriminating power. And the latter will prevail. Our agreement is due much less to the similarity of objects than to the uniform way in which we react to their presence. Conceptualisation will undoubtedly be an effort of objectivity, but on average it will develop in an unexpected sense: in fact, the object is not able to invoke the purification of the concept, as its needs are always minimal since at the very least a single feature would be enough to designate it: instead it is the spirit that projects multiple schemes, a geometry, a construction method and even a rectification method. This last aspect translates the need for novelty, for creation, which is undoubtedly a spiritual need, no less essential than assimilation. Conceptualisation, in its final form, is the search for an end. In fact, if conceptualisation is examined at the end of Duhamel's ternary process (comparison, abstraction, generalisation), an authentic teleological force is captured in it when it returns to reality as a general voluntary form applied to a new subject. A concept strives towards generalisation. To do this, it will reproduce itself into multiple domains, going so far as to rectify its data in some aspect. Speculative thinking has a tendency to become normative."

The quotation above allows us to better understand how Bachelard fully grasped the Galilean duality of the progress of scientific knowledge, while he also realized that thought, by its intrinsic nature, cannot be reduced to a general and abstract scheme (as empiricism would do instead), because, on the contrary, it always lives and develops within a precise *dynamic conceptual network*: "Thought begins only with a verb, and is contemporary with the connection between concepts". Seen in this perspective, "synthetic judgment is necessarily a creator, but it must be so progressively, by slow assimilation". Science, therefore, walks with a "sailor's gait", relying on both the conceptual and the experimental dimensions: "in its first momentum it is a discovery full of uncertainty and doubt. Cautious judgments are at its roots; verified cases are its successes." A success that often "fossilises" the act of knowing in a consolidated mechanism whose true nature always springs, however, precisely and only from that tension and that cautiousness by which a concept, passing through doubt and uncertainty, builds knowledge *in fieri*, which is always approximate and always correctable, because research, as Popper also said, is always open and endless. Of course, we should not overlook the difference between Popper and Bachelard regarding this intrinsic "openness" of research. For Popper, "openness" is rooted in its own radical conventionalism, *à la* Xenophanes,[13] by virtue of which all human knowledge would be nothing more than an extremely large web of "conjectures" that at best can be partially "corroborated", until they can be finally falsified. The Popperian "openness" of research therefore refers to his exquisite cemeterial conception of history.[14] On the contrary, the "openness" of which Bachelard speaks is a 'plant openness' which benefits from continuous rectification, precisely because it constitutes a path of continuous and equally tenacious growth, thanks to which humanity is actually able to delineate a technical-scientific heritage of knowledge and of operating practices. His, as we have seen, is also a *teleological* openness which has a profoundly different meaning from the Popperian one, because it does not imply at all a leap from a falsified theory to a forthcoming "corroborated" theory, to be falsified in the near future, but rather implies a continuous adjustment of growth and construction which, thanks to an infinite succession of continuous approximations, allows us to conceptually assimilate the world into an increasingly objectified reality, although we are never able to grasp the real world in an exhaustively metaphysical way.

Starting from this innovative, intrinsically dynamic image of scientific knowledge, Bachelard always recognised his theories in a form of open, dynamic and "supra-rationalist" rationalism. In fact, for Bachelard it is necessary to have the ability to abandon the traditional form of "closed rationalism", typical of the metaphysical tradition which, especially in the modern age, has forged the great rationalist reflection of authors such as Descartes and Spinoza, to mention only two emblematic names, to make room for an "open rationalism": "the happily unfulfilled reason can no longer fall asleep in a tradition; it can no longer rely on memory to recite its tautologies. We should challenge reason and challenge ourselves tirelessly. Reason is in combat with others, but first of all with itself. This

---

[13] In this regard, however, it is worthwhile to refer to the beautiful chapter dedicated by Popper to Xenophanes of Colophon, which can be read in his posthumous volume *The world of Parmenides* (an authentic and extraordinary "Essay on the Presocratic Enlightenment" as the subtitle of the English edition rightly describes it. Conversely the Italian publisher, with the opposition of the translator, opted for a vaguer description: "Discovery of the Presocratic Philosophy"), edited by Arne F. Petersen, with the assistance of Jorgen Mejer, Routledge, London, 1998. The title of the chapter is already emblematic: *The unknown Xenophanes: an attempt to establish his greatness.*

[14] In this regard, see Fabio Minazzi, *Riflessioni critiche sulla filosofia di Popper*, "Epistemologia", XIII, 1990, pp. 221–236, as well as my monograph *Il flauto di Popper*, Franco Angeli, Milan, 1994.

time, it has some guarantee of being incisive and young."[15] From this point of view, for Bachelard science is "one of the most irrefutable evidence of the essentially progressive existence of thinking beings. Thinking beings think thoughts which try to know. They do not think existence." In fact, "thoughts which try to know" means thoughts which strive to understand reality conceptually by objectivising it, while thoughts that are supposed to be able to conceive existence are ingrained in the traditional metaphysical ontology that from Parmenides to Heidegger claimed that it was able to grasp the concept of Being as such. Against this metaphysical ontology he referred to the "permanent rationalism" which distinguishes almost the whole western tradition, underlining the "dialectical condition" of scientific thought. To illustrate the "dialectical" nature of scientific thought, Bachelard started from the classical *Études galiléennes* by Alexandre Koyré (originally published in Paris, Hermann, 1966), which allowed him to grasp a double movement—ascending and descending—present within scientific reflection and its intrinsic dynamism. In general, the empiricist tradition has always emphasised the decisive role of the experimental verification of theoretical propositions. However, Bachelard observed, relying on some subtle reflections by Koyrè, that there is also an inverse movement which is proper and typical of modern thought since "it is necessary [...] that a fact, to be truly scientific, is *theoretically verified*". The experimental verification of theoretical statements is therefore not enough, because the latter must necessarily intertwine with the theoretical verification of facts themselves. But what can this theoretical verification be if not the conceptualisation through which a certain and partial aspect of reality is made meaningful within a specific theory with which, precisely, the world can be thought in order to be known? In this way, for Bachelard a specific dialectic, always open and progressive, of scientific thought is implemented which is thus able to constantly intertwine the pole of theory with that of experimentation, putting in place a complex movement of thought and action. Naturally, the open rationalism theorised by Bachelard was a kind of rationalism that tended towards improvement, precisely to experience the actuality of its time, which finds in science its inevitable reference point for progression. In the famous discussion *On the Nature of Rationalism* promoted by the "Société française de Philosophie", in the session of Saturday 25 March 1950, with an extensive report by Bachelard, he had the opportunity to return to the specific nature of his applied rationalism and stressed again that the truth of his "suprarationalism" had its roots "in the work of experience through rational activity" precisely because his is a

---

[15]Translated from the Italian edition of Gaston Bachelard, *L'impegno razionalista*. Preface by Georges Canguilhem, edited by Francesca Bonicalzi, Jaca Book, Milan 2003, p. 29, while the quotations that appear later in the text are taken, respectively, from the following pages: p. 54; p. 60 (italics in the text); p. 75; p. 149.

"rationalism at work". A rationalism that lives on the dialectic of thought itself, which is realised in the scientific work within which the "specialisation" itself can only expand and enrich the spirit of research and our own reflections. For this reason, for Bachelard "to be a rationalist, one has to go and look for [...] rationalism where it is: in scientific thought". It is precisely the analytical study of the technical-scientific heritage, captured in all its intrinsic articulation, that allows us to develop a Bachelardian "regional rationalism" which is deeply in accord with the identification of the different "regional ontologies" already identified and thematised by Husserl in his classical and emblematic phenomenological recognition of knowledge entrusted to the pages of its first *Logical Investigations*.

So if in the sciences (physics, chemistry, but not only, *of course!*) the rational organisation and experimental experience are always intertwined and in critical "constant cooperation", as Galilei observed, it is then inevitable to note that in Bachelard's reflection the history of science cannot fail to acquire a primarily privileged and decisive position. Why? For the simple but decisive reason that "a scientific truth is a truth understood. A true idea, understood as such, cannot be turned into a false one. The temporality of science is an increase in the number of truths, an improvement of the depth of the coherence between truths. The history of the sciences is the story of their growth and development." For this underlying reason in Bachelard's opinion the decline of civilisation is fundamentally alien to the spirit of the history of science, precisely because "the history of science is always described as the history of a progress of knowledge. Readers moves from a state where we knew less to one where we know more. To think historically about scientific knowledge translates into the description from less to more. It is never the reverse: from more to less. In other words, the cornerstone of the history of science is clearly oriented towards a better understanding and a wider experience." For this reason, the history of science must fulfil some obligations which do not apply for those who deal with historical research as such. A historian must in fact be exempt from expressing a judgment, because if anything, he must help us understand the reasons that account for a specific historical situation in its own dynamic. On the contrary, for Bachelard, a science historian should always be able to make "value judgments": "the history of science is at least a fabric of implicit judgments on the value of scientific thought and discoveries. A science historian, who clearly explains the value of each new thought development, helps us to understand the history of sciences." For this reason, the history of science can only be an *assessed* history, "assessed in the details of its development, with a meaning that must be continually refined through the values of truth." A science historian should thus be able to highlight "the lines of progress" in his documents. Naturally, in order to produce an evaluation of the past, science historians cannot exempt themselves from competence in sciences in actuality: "in order to evaluate the past, science historians must know the present; they should learn, as best they can, the science on whose history they want to report. From this point of view, the history of sciences is therefore strictly connected "to the actuality of science". In this way, Bachelard's epistemology thus manages to underline a profound and essential link between the history of science and epistemological reflection by promoting a critical awareness of scientific knowledge, which is instead almost absent from the tradition of empiricist-verificationist epistemology (as well as from epistemology inspired by falsificationism.)[16]

7   For a historical-critical epistemology

7.1   The critical split between absoluteness and knowledge

Leaving behind the very synthetic and elliptical overview expounded in the previous paragraphs, we can try to outline the possible features of a future "historical-critical epistemology". Firstly, a decisive aspect has emerged which directly concerns the most rigorous and correct idea that can be delineated of science and human knowledge itself. From Banfi's and from Bachelard's approach the discovery of the *conceptual dimension* of science has emerged, although by following two completely different and autonomous research paths. When we speak of the "conceptual dimension" of science we mean to emphasise how science operates and is develops through its own *style of thought*, which constitutes its fundamental core. In other words, the "conceptual dimension" of science coincides precisely with "scientific thought" and science is such, *in primis* and *ante omnia*, because it produces thought, i.e., *scientific thought*. This observation, which emerges forcefully from both the Banfian and Bachelardian traditions, naturally finds its precise derivation, both theoretical and historical at the same time, in the Kantian discovery of the transcendental, which enabled Kant to elaborate his famous "Copernican revolution" by virtue of which Kantian criticism was able to initiate a plastic and articulated examination of human reason. Not only that: the beginning of criticism also coincided with the radical challenging of every metaphysical claim, since the Kantian transcendental

---

[16]Certainly Imre Lakatos, with his sophisticated falsificationism, underlined the close (*Kantian*) interconnection between the history of science and the philosophy of science: the latter without the former is empty, while the former without the latter is *blind*. However, Lakatos, with his methodology of scientific research programs, is also wants to carry out a "rational reconstruction" of the history of science *in the text*, relegating *to the notes* the real and effective history, in order to show how the latter would have "misbehaved" with respect to rational reconstruction! (See I. Lakatos, *The Methodology of Scientific Research Programmes*, edited by John Worral and Gregory Currie, Cambridge University Press, Cambridge, 1978, pp. 118–120). In this way in the Lakatosian reflection the typical theoreticism derived from Popper's teachings (see F. Minazzi, *Il flauto di Popper*, op. cit.) which epistemologically engulfs the history of science...

denied a basis to metaphysical ontologism (from Parmenides to Heidegger), i.e., the claim of being able to understand Being as such. *Against* this recurrent and traditional ontological-metaphysical temptation, the Kantian transcendental turn highlights, however, how human beings can never have direct and immediate access to reality, since the latter can only be grasped and known in an ever partial and delimited way. In this perspective, Kantian criticism, relying on the genesis of modern science which undoubtedly constituted a decisive *turning point* in the history of modernity, introduced the notion of *scientific objectivity* determining a development of undoubtedly historical significance. In fact, the "Copernican revolution" (which we could also identify as authentic "Kantian revolution"), to put it in Jules Vuillemin's words, led to a real split between knowledge and the absolute, a break that Kant generated without denying authentic cognitive scope to human scientific knowledge, thus preserving a precise and determined sense to the question of the difference that exists between reality and appearance, between what is necessary and what is instead contingent and did so within a philosophy which precluded the possibility of talking about things in themselves. But in this respect, it is best to allow Vuillemin to speak for himself:

> Si pensée physique et théorie de la connaissance ne font qu'un chez Kant, celle-là éclairera la nouveauté révolutionnaire de celle-ci. Avant Kant, la philosophie classique essaie, une fois ébranlés les systèmes théologiques du Moyen Age, de découvrir un absolu susceptible de fonder la vérité. Par exemple, les concepts de substance, de cause, de force, de nécessité reçoivent ce rôle de substituts de Dieu. L'acte révolutionnaire de Kant dans l'histoire de la pensée, sa "révolution copernicienne", a consisté, en reprenant l'analyse de ces différentes notions par rapport à la fonction qu'elles exercent dans la connaissance objective, à montrer que, loin de monnayer l'absolu, elles ne conservaient de signification que dans les limites de l'expérience possible, c'est-à-dire si on les coupait de leur contexte théologique. A cet égard, la théorie kantienne de la connaissance est la première théorie conséquente et vraiment philosophique d'une connaissance sans Dieu.[17]

This Kantian philosophical theory of knowledge that no longer needed to anchor itself to the notion of divine absoluteness, also freed science from any undue reference to the dimension of absoluteness. Naturally the post-Kantian reflection variously elaborated, misinterpreted and even openly fought and rejected the Copernican approach outlined by Kant, so much so that his own philosophical lesson, decidedly anti-metaphysical, often ended

---

[17] J. Vuillement, *Physique et métaphysique kantiennes*, Presses Universitaires de France, Paris 1987[2], p. 358.

up by being an almost exclusive property, precisely and paradoxically, of the metaphysical tradition itself. Which also led to the considerable decidedly anti-Kantian hatred of most of the exponents of the *Wiener Kreis*, who in relation to a Kantian philosophy, at the time almost the exclusive prerogative of metaphysicians, then certainly "threw out the baby with the bathwater", completely disregarding the Kantian epistemological approach that also emerged, with strength and equal fruitfulness, in Marburg's neo-Kantian tradition, which was probably expressed at its best and in the most original way by Ernst Cassirer's critical and constructive analysis. In any case, the problem posed by Kant, insofar as it captured a decisive aspect of human knowledge, could not fail to re-emerge also in the later reflection that was often constructed, as happened for example in the case of Bachelard, autonomously and independently of Kant's teachings. In any case, the problem encountered the Kantian epistemology could not fail to re-emerge in the reflection following his works. And this actually happened to the extent that during the twentieth century the conceptual dimension intrinsic to the scientific enterprise was strongly emphasised. Naturally this recognition of the presence of scientific thought, its relevance and its heuristic function were not recognised by everyone because the other traditions of thought, still rooted in traditional metaphysical ontology, openly fought against this perspective, as happened, for example, with the reflection of Martin Heidegger for whom, as is known, "die Wissenschaft denkt nicht." In the twentieth century we were thus faced with two different horizons of thought: on the one hand there were those who thought that science is essentially based on the ability to produce its specific decidedly innovative knowledge, fundamental for the human understanding of the world and, on the other hand, there are those who denied this possibility and who opposed traditional metaphysics to scientific thought and to the development of technology, claiming the use of thought as such as an exclusive privilege of traditional metaphysics. Which then is also found in the common sense that pervades our societies, if it is true, as it is true, that generally the scientific dimension is perceived as an eminently "technical" structure which generally denies any specific cultural value, while the meaning of "culture" is arbitrarily restricted to the world of humanistic research only. Which brings us back to the dramatic split between the so-called "two cultures" by which the fruitful link between science and philosophy, which has always existed in the long-term history of Western tradition, is undoubtedly undermined and neglected, to affirm an absolute "split" between the two, which, in part, has been recorded only in the last three centuries of western history.[18] But instead of critically investigating and studying the profound

---

[18]For a serious and systematic critical reflection on the connection between the "two cultures" within western tradition, a reference to the acute volume by Giulio Preti still

and intrinsic reasons for this alleged incompatibility, this "split" is instead exhibited and assumed in a rather partial way and it is often presented instrumentally, as an element that should precisely play exclusively in favour of the humanistic tradition, which allegedly is the only one capable of producing thought. Furthermore, education systems and trainings contribute to the maintenance and social diffusion of this profound distortion of the cultural dimension, which often and willingly insists on presenting the "two cultures" as divided and armed against each other. In the educational field, this split is fuelled by the very way in which the humanities and scientific disciplines are studied: for the former, a decidedly historical approach is used, while for the latter, a decidedly and deliberately ahistorical approach is employed. In this way the school system—from primary schools to universities *included*—does nothing but reinforcing the split between the "two cultures", preventing us from understanding the fruitful connections that have always nourished the relationship between scientific and philosophical thought.[19] Why? Precisely because scientific knowledge (mathematics, geometry, physics, natural sciences, astronomy, etc.) are taught in a strictly ahistorical way, insisting only on the "technical-algorithmic" aspect, (systematically) neglecting the conceptual dimension of science. On the other hand, the humanities are taught adopting a tendentially historical approach which, however, inevitably weakens them at least to the extent that in our schools there is an increasingly widespread "particulate" teaching based on purely technical education and purely technical training, which no longer educates, but is limited to instructing, neglecting a cultural formation worthy of its name.[20]

7.2   A new unitary image of human knowledge

Secondly, this epistemological approach, which, as we have seen, fully highlights the *conceptual dimension* of science, must then lead us to review, *ab*

---

remains fundamental. *Retorica e logica. Le due culture*, Einaudi, Turin, 1968, now available in the new amended and enriched edition, edited with the introduction and notes by Fabio Minazzi, Bompiani, Milan, 2018.

[19] The only work, on an international level, that tried to openly combat this avowedly dichotomous approach to culture was the one promoted and largely written by Ludovico Geymonat with the publication of his monumental *Storia del pensiero filosofico e scientifico* (*History of Philosophical and Scientific Thought*), Garzanti, Milan, 1970-1976, 7 vols., Which is still today, worldwide, the only work that endeavored to illustrate the constant and always fruitful link between philosophical thought and scientific thought through the entire course of the history of the western tradition.

[20] In this regard, see the proceedings of a conference specially dedicated to *La scuola dell'ignoranza* (*The school of ignorance*), edited by Sergio Coltella, Dario Generali and Fabio Minazzi, Mimesis, Milan-Udine 2019), which offers a mercilessly critical examination of the overall degradation of education in Italian schools, which fully mirrors the parallel overall degradation of Italian universities following the reforms of various Ministers for Education (Berlinguer, Moratti and Gelmini.)

imis fundamentis, the very nature of *human knowledge*. Which can happen at least in a double critical sense. In the first place, in fact, it is necessary to critically distance the dimension of knowledge from the horizon of absoluteness, by elaborating a *new conception of the objectivity of scientific knowledge*. This, for example, is the path followed by an epistemologist like Evandro Agazzi who in his most recent volume, *Scientific Objectivity and Its Context*,[21] addressed the objectivity of scientific knowledge by systematically referring to its different constitutive contexts. In this way, Agazzi's proposal once again allows us to separate the objectivity of knowledge from the dimension of absoluteness, recovering a notion of knowledge that turns out to be true, absolutely true, only within defined, strictly circumscribed areas. In this perspective, scientific knowledge is therefore certainly "relative" knowledge, but it is such only and exclusively within a limited and finite sphere of objectification of the world. Within each cognitive context there is therefore a sort of critical convergence between absoluteness—which allows us, in fact, to distinguish what is actually known, in a correct way, from what does not instead constitute knowledge and is configured, therefore, as an "error" that must necessarily be corrected—and the very relativity of knowledge, which is such precisely because it refers to a limited and circumscribed area.[22]

The affirmation of the critical construction of the objectivity of scientific knowledge within its specific contexts also allows us to profoundly modify our overall image of human knowledge. In fact, this can no longer be associated solely and exclusively with the scientific dimension because it is instead necessary to elaborate a much richer, more articulated, plastic and comprehensive image of human knowledge as such. In fact, it cannot be denied that there is knowledge also within traditional "humanistic" fields. For instance, Lorenzo Valla published, in 1440, the *De falso credita et ementita Constantini donatione declamatio*, demonstrating in a *philologically* rigorous way, that the so-called "Donation of Constantine" traditionally exhibited by the Catholic Church to justify its temporal power, was, in reality a "historical forgery". Well, can this writing be regarded as an example of knowledge or not? Historically, the cognitive contribution of Valla's text cannot be seriously denied, even if in this case it is a predominantly negative cognitive contribution, precisely because the Catholic Church itself, *after* the publication of Valla's work avoided again showing the presumed "Donation of Constantine" as indisputable proof to justify its illegitimately

---

[21] Springer, Cham Heidelberg New York Dordrecht London 2014, Italian translation by Giovanni Carrozzini, Elisabetta Scolozzi and Giulia Santi, with editorial revision and final editing by Fabio Minazzi, promoted by *Centro Internazionale Insubrico "C. Cattane" and "G. Preti"* of the University of Insubria, published by Bompiani, Milan, 2018.

[22] On this issue see F. Minazzi, *La riflessione filosofica di Evandro Agazzi*, "Giornale di Metafisica", year XL, 2/2018, pp. 732-737.

exercised temporal power. However, this philological knowledge was built using a methodology and criteria that are profoundly different from those used in the natural sciences. This must then lead us to elaborate a new and more articulated critical image of human knowledge. To do this we can employ the Husserlian suggestion by which each discipline constitutes its own specific "regional ontology". Moreover, Husserl's reflection is valuable because it implied a pluralisation of the traditional Kantian concept of the transcendental. If in fact for Kant the transcendental was rooted in the only form of scientific knowledge actually available at the time, i.e., Newtonian physics, on the contrary the increase itself of the contemporary scientific heritage and its increasingly rapid differentiation and articulation allow the pluralization of the horizons of transcendentality, understanding the different levels within which this form of critical meta-reflection on human knowledge can be exercised. In other words, it is necessary to know how to rethink human knowledge in a unitary and, at the same time, very articulated way, in order not to sacrifice all its critical potential without, however, renouncing to provide an overall picture of human knowledge that is built differently within the different areas of cognitive research. In this perspective, in short, we must definitely turn our backs on that tradition that determined a sort of cultural "imperialism" of physics-mathematics that led us to consider research as "scientific" only and exclusively to the extent that it is mathematised. It is necessary to consider how this traditional conception of epistemology actually resulted in a cultural hegemony which, for example, induced a thinker like Kant to argue that a discipline is all the more "scientific" the more it can be "mathematised" with the good result that the more mathematics is present within a discipline, then the more this discipline was "scientific". This approach then explains why the so-called sciences have been distinguished between "hard" and "soft" sciences using a quasi-"pre-Northern League" lexicon that helps us better understand the possible epistemological deformations that this conception of knowledge can inevitably feed. The alleged ideological rift between the "two cultures" finds in fact its key element in this mathematical approach, in the name of which it then claims to hierarchise knowledge as such, placing the hard sciences at its top and then relegating to increasingly lower levels the other disciplines that cannot be mathematized on an equal formal plane. In this way, mathematics, from a heuristic tool that has increased the physical investigation of the world, risks turning into an engulfing epistemological bond, in the name of which the very attribute "scientific" can be given or denied. Against this dogmatic model of knowledge, it is therefore necessary to elaborate a much more articulated and plastic vision of human knowledge that can actually proceed through different paths by devising different "regional ontologies", or rather different cognitive "regions" which are established by inserting themselves into different conceptual traditions, which have elaborated different conceptual tools, specific verification (and falsification) methods, giving rise to specific problems and also to a peculiar tradition able to solve given problems. From this point of view it is then necessary to rework, also in this case *ab imis fundamentis*, the idea itself of human knowledge, referring back the exceptional lesson of Leonardo da Vinci who, not surprisingly, already anticipated, at the dawn of modernity, the power and the strategically decisive fascination of developing another and different unitary culture, able to displace the reifying unilaterality of both scientific and humanistic culture, in order to outline a new and alternative, much richer, more articulated and fruitful cultural synthesis. As Banfi wrote, in his essay on *The humanity of Leonardo da Vinci*, the genius of Da Vinci consisted in fact in being aware of the profound harmony that exists between humanity and nature because in his conception of culture Neoplatonism is stripped of all its traditional idealistic elements in order to re-emerge in its first source, that of Greek naturalism,

> purified by the sense of the unity of nature in which human beings live, which is humanity itself. Well, to understand this reality is to recognise this reality. This is the great task of humanity, this is the extraordinary development initiated by Leonardo and it is possible to understand it because we are made of the same substance, because the vibrations, which are present in nature, exist also in human beings, because macrocosm and microcosm correspond to each other and there are two ways to proceed: one is art, the other is science. These are the two ways that make it possible for humanity to discover and to conquer reality.[23]

7.3   A new image of the historicity of human knowledge

Thirdly, these considerations forcefully pose the problem of always taking into due account that human beings always live within a specific tradition since they can reason, speak, think and elaborate their own speeches and actions, but they always perform these actions and thoughts in a concrete historical context from which they can never prescind themselves. The same philosophical reflection is fuelled—as indeed happens to science—precisely by this paradoxicality: it aspires to a universal and necessary knowledge, therefore able to ignore the historical concreteness, but to do so it must always start from the particular historical context in which individuals find themselves living, thinking, reasoning and acting. If this historical dimension of tradition is considered, culture, experience and knowledge themselves are transformed into something abstract and arbitrary, precisely because they are considered outside the precise historical contexts that produce and

---

[23] A. Banfi, *Scritti letterari*, edited by Carlo Cordié, Editori Riuniti, Rome, 1970, p. 82.

substantiate them, making them flesh of our flesh and blood of our blood. In fact, languages, problems, categories to which we can refer to reason, reflect and live, do not exist, in Antonio Labriola's words, as "caciocavallo cheeses hung in a deli", but they are born and are always transmitted by human beings through a complex and articulated historical process, through which the languages, the categories, the problems to which we can possibly and primarily dedicate ourselves reach us. Consequently, philosophy has the task of critically reflecting on all these different forms and structures of the various traditions, and also of making explicit all the contradictions as well as their possible various divergences. In this specific perspective it is possible to outline an original form of historical-critical-objective transcendentalism with a logical neo-realist structure such as that theorized by Giulio Preti. In the last phase of his more mature reflection, in fact, he thought that he could delineate

> a historical-objective transcendentalism, which surveys the constructive forms of the various universes of discourse through a historical-critical analysis of the ideal languages that serve as models for these universes, from the rules of method that have been imposed historically and still apply in knowledge, etc. In short, it is a transcendental Ontology (or rather transcendental ontologies) which does not claim to understand the forms and structures of a Being in itself, but seeks to determine the way (or ways) in which the category of being is enacted in the historically mobile and logically conventional (arbitrary) construction of the ontological regions by scientific knowledge (in particular) and culture (in general).[24]

Philosophically speaking, this attitude highlights, once again, the exquisitely critical meta-reflective character of the philosophical activity which ultimately, following *Kant and Husserl*, investigates, first of all, the historical configuration of a tradition assuming it in its actual concreteness, and then develops a reflection that never seeks to unravel the Being of the world, because, more modestly, it limits itself instead to investigate, critically, the various constitutive structures of the different universes of discourse, in order to reconstruct the historical mobility of human knowledge. Certainly, in this perspective the intrinsic *relativity* of human knowledge is clearly perceptible, since

> whatever is based on historical experience passes away with that experience. In becoming aware of the relativity of all scientific development, epistemology, which is itself a scientific construct, becomes aware of its own relativity. From the logical point of view, there is no difficulty. Having realised that the notion of the 'eternally true'

---

[24]G. Preti, *Philosophical Essays*, vol. I, op. cit., p. 297, while the quotation that follows in the text is taken from p. 70, again in the first volume of this work.

is meaningless, only the notion of 'historically [hence relatively] true' has a meaning, and this applies to all knowledge. The difficulty is psychological: epistemology so conceived offers no hope to those who yearn for the eternal, those who see Reason as a factory whose job is to turn out goods that will appease the yearning for eternity, for eternal truth and certainty. But this is the 'defect' of all forms of culture that have raised humanity out of barbarism. For those who do not have this yearning, for those who tranquilly accept the possibility of dying, in the fullest sense of the word, but who also lay great importance on forming the clearest ideas possible, the most intersubjective possible, ideas that help to release us and our fellow men and women from nightmares and phantoms of the afterlife and make earthly humanity's house as comfortable and pleasant as possible—for such people the 'defect' is transformed into the highest value.

In this precise hermeneutic and critical-epistemological key, the intrinsic relativity of human knowledge thus becomes a pivotal point by which it is possible to actually construct a critically more appropriate image of knowledge by referring it, precisely, to that "defect" of "relativity" that historically "raised humanity out of barbarism." Which, in fact, coincides with the actual history of humanity. But when philosophical reflection comes into play, all these levels inevitably become complicated and distinctions must be made, precisely because there are different degrees of reflection and thought. In fact, there is a reflection—which Banfi calls "pragmatic reflection" which generally constitutes a first reflexive and thoughtful reworking of some particular sectors of human experience. At this first level, thoughts produced by pragmatic reflection inevitably undergo all the constraints of a pragmatic reflection that struggles to detach itself from the horizon of life experienced in its practical-sensitive activity. Philosophical reflection instead rises above this level, freeing itself from pragmatic interests in order to investigate the transcendental laws of constitution and also of intrinsic movement of the pragmatic forms of reflection themselves. In this progressive detachment from the pragmatic horizon, different levels can thus be identified, from that of the *moral philosophy* (which systematises and organises the values within which the proper and specific action of the world of praxis takes place) to an even more abstract and higher level, in which a *philosophy of morality* is conceived, which carries out a critical meta-reflection on the universe of discourse of moral philosophy, on its categories and its constitutive structures.[25]

---

[25]For a systematic study of all these issues, however, see A. Banfi, *La ricercar della realtà*, edited by Guido Davide Neri and Gabriele Scaramuzza, with the collaboration of Barbara Cavaleri, Istituto Antonio Banfi-Società Editrice il Mulino, Bologna, 1996, 2 vols., vol. II, with particular reference to the second part *La vita della cultura*, pp. 363–721 and the fundamental essay of 1934, *Sui principi di una filosofia della morale*, pp. 493–558.

In any case, from this perspective, philosophy always takes the form of a reflection on culture, whose *materia subjecta* is never experience (or reality) as such, but the different and multiple cultural forms in which experience (or reality itself) is thought, understood, felt, lived, etc. In this perspective, in Jacques Ruytinx's words, author of *La problématique philosophique de l'unité de la science*, "la philosophie est une *métaréflexion dont le niveau est toujours susceptible d'être déplacé.*"[26] Within this framework, according to which philosophy "advances" only to the extent that it "steps backward", one can distinguish different specific levels specific to the philosophy of science as such. We can thus identify a first level of reflection on science which coincides with a methodological one, which on the one hand can only bend itself critically on the different ways in which each disciplinary scientific field is constituted, while, on the other hand, it can also try to detach itself from this level to reflect on the logic of scientific discourse, by specifying the logical conditions of scientific nature itself. Rising to this more general level of epistemological reflection then leads to a sort of reflection on the "logic of science" which for a large part of contemporary epistemologists coincides, *de facto* with the philosophy of science *tout court*.[27] But on the other hand also this investigation on the "logic of science" tends to become increasingly specialised, transforming itself, in turn, into a sort of technical and scientific discipline in relation to which philosophical reflection can react by rising to a level of greater critical generality that considers the previous level as its own *materia subjecta* in order to build a more general and decidedly more philosophical reflection. Some authors then tend to distinguish these two levels by talking about epistemology for the second level that investigates the "logic of science" and instead referring to philosophy of science for the third level that investigates the nature of science in its most extensive structural generality.[28]

---

[26] J. Ruytinx, *La Problématique philosophique de l'unité de la science: étude critique*, Le Belles Lettres, Paris 1962, p. 339, note no. 2, italics are in the text.

[27] A good model of this decidedly specialised conception of science as such can be found, for example, in the excellent *Springer Handbook of Model-Based Science*, Lorenzo Magnani, Tommaso Bertolotti eds., Springer, Dordrecht Heidelberg London New York 2017, in which a conspicuous and articulated number of specialists tackles a multiplicity of different, somewhat narrowly delimited themes, with a deliberately technical and specialised language, which seems however, to exclude a possible and different exquisitely philosophical evaluation of the object of their reflections. In this case, epistemology is transformed into a highly technical and specialised discipline that has nothing to envy to the specialisation of other scientific disciplines, even if at times it seems almost that, at least in some more technical and deliberately specialised contributions, the philosophical dimension finally risks, paradoxically, disappearing...

[28] A critical-systematic reflection on all these different levels of philosophical investigation of science was moreover developed, with the usual acuteness, by Preti in the introductory part of his excellent *Lezioni di filosofia della scienza* (1965–1966), edited by Fabio Minazzi, (Franco Angeli, Milan 1989, pp. 53–61) to which I directly refer.

To these reflections must be added the further consideration that philosophical reflection itself, at least to the extent that it wants to be configured as a proper and specific reflection of a scientific philosophy, finds in the philosophy of science its strategic and emblematic point of reference, so much so that in authors such as Hume, Kant and the neo-positivists themselves, the philosophy of science ended up by identifying itself, not by chance, with the same gnoseology intended, precisely, as general philosophy. Not to mention that science itself, in turn, can naturally be subject to different meta-reflective considerations, because it is configured both as *knowledge* (although, as we have seen, it is then questionable whether it is the only possible form of knowledge), both as a historical element of civilisation (precisely: the civilisation of sciences!) and as a peculiar discipline which is exactly studied and investigated, in its historical actuality, by the philosophy of science as such. The whole plurality of these multiple levels of philosophical investigation of science can and must always be traced back to the peculiar meta-reflective character of philosophical thinking as such. If we do not do it, as often happens today in the international epistemological debate, we will inevitably end up by losing sight of both the specific and the intrinsic cultural value of science (and, consequently, of philosophy of science itself), and also of its distinct theoretical importance as well as its value in the history of civilisation.

www.ingramcontent.com/pod-product-compliance
Lightning Source LLC
Chambersburg PA
CBHW071944100426
42737CB00046BA/2282